Bioethics in Law

BIOETHICS IN LAW

By

Bethany J. Spielman, JD, PhD

Department of Medical Humanities
School of Medicine and School of Law
Southern Illinois University, Springfield, IL

HUMANA PRESS ✳ TOTOWA, NEW JERSEY

© 2007 Humana Press Inc.
999 Riverview Drive, Suite 208
Totowa, New Jersey 07512

humanapress.com

Due diligence has been taken by the publishers, editors, and authors of this book to assure the accuracy of the information published and to describe generally accepted practices. The contributors herein have carefully checked to ensure that the drug selections and dosages set forth in this text are accurate and in accord with the standards accepted at the time of publication. Notwithstanding, as new research, changes in government regulations, and knowledge from clinical experience relating to drug therapy and drug reactions constantly occurs, the reader is advised to check the product information provided by the manufacturer of each drug for any change in dosages or for additional warnings and contraindications. This is of utmost importance when the recommended drug herein is a new or infrequently used drug. It is the responsibility of the treating physician to determine dosages and treatment strategies for individual patients. Further it is the responsibility of the health care provider to ascertain the Food and Drug Administration status of each drug or device used in their clinical practice. The publisher, editors, and authors are not responsible for errors or omissions or for any consequences from the application of the information presented in this book and make no warranty, express or implied, with respect to the contents in this publication.

This publication is printed on acid-free paper. ∞

ANSI Z39.48-1984 (American Standards Institute) Permanence of Paper for Printed Library Materials.

Production Editor: Melissa Caravella

Cover design by Patricia F. Cleary

For additional copies, pricing for bulk purchases, and/or information about other Humana titles, contact Humana at the above address or at any of the following numbers: Tel.: 973-256-1699; Fax: 973-256-8341; E-mail: orders@humanapr.com; or visit our website at www.humanapress.com.

Printed in the United States of America. 10 9 8 7 6 5 4 3 2 1
eISBN 1-59745-295-5

Library of Congress Cataloging-in-Publication Data

Spielman, Bethany, 1952-
  Bioethics in law / by Bethany J. Spielman.
    p. cm.
  Includes bibliographical references and index.
  ISBN-13: 978-1-58829-434-0 (alk. paper)
  ISBN-10: 1-58829-434-X (alk. paper)
 1. Medical laws and legislation--United States. 2. Judicial process--United States. 3. Medical jurisprudence--United States. 4. Bioethics--United States. I. Title.
  KF3821.S687 2006
  344.7304'1--dc22
                            2006012286

Dedication

For Keith

Preface

The idea for *Bioethics in Law* began more than a decade ago, while I was studying social science and law. I was particularly interested in the collaborations that comprised social science *in* law. Economic and social data in the pioneering Brandeis brief had been used to defend an early 20th-century labor law; surveys of consumer confusion had helped resolve trademark infringement cases; psychologists' predictions of future violence had informed capital sentencing decisions. Additionally, Kenneth Clark's "doll studies," cited by the Supreme Court in *Brown* v. *Board of Education*, had helped change the course of American history.[1]

During that time, however, I was most intensely interested in bioethics, a relatively young field whose relationships to law had not been well analyzed. I wondered whether there could or should be a bioethics *in* law, because bioethics, unlike the social sciences, was not only in its infancy, but also had distinctly normative features, which might not mesh easily with law's own normativity.

Bioethics commission reports were appearing; bioethicists were starting to testify as experts; the *Karen Ann Quinlan* court had decided that ethics committees, rather than courts, should make decisions about forgoing life-sustaining treatment. Legal scholar George Annas would occasionally analyze the contributions of bioethics to law in a column in the *Hastings Center Report*. For the most part, however, bioethicists, who were pleased to have professional opportunities outside the classroom, did not engage in analysis of how and why their contributions were used—or not used—in law. Their participation as expert witnesses, as well as in health care ethics committees, institutional review boards, and bioethics commissions seemed to be enough; as one prominent bioethicist put it, they were now "players."

For their part, legal scholars were more interested in the legal implications of medical technologies, such as organ transplantation and life-sustaining treatments, than they were in how that law had been informed—or not been informed—by resources from the field of bioethics.

Bioethicists have since broadened their professional engagement with the world and grappled with more recent developments including pharmacogenetics, biobanking, and nanotechnology. Legal scholarship on these issues, as well as scholarship on "law and norms," has flourished. Scholars of social science in law have broadened and deepened their inquiry and even judges now acknowledge that the proposition relied on in *Brown* v. *Board of Education* was a product of normative judgment rather than Clark's studies. However, analysis of bioethics' input to the legal system is scant and remains undeveloped.

This book begins to develop an analysis of bioethics in law. It expands on an approach used in an earlier work. That approach was characterized as one of two key directions for the future of scholarship in bioethics and law: the "law of bioethics."[2] Bioethics in law as illustrated here involves applying legal norms to bioethics, as would a "law of bioethics" approach; but preceded by steps such as receiving and assessing bioethics resources.

The focus of analysis here is bioethics as it has come to law during the last decade. This is not bioethics in an abstract or idealized sense, but the bioethics in actual communications that have found their way to law: health care ethic committee recommendations, institutional review board determinations, bioethics commission reports, bioethics research notes, briefs of bioethics amicae curiae, and bioethics expert testimony. Although *Bioethics in Law* touches on traditional legal and bioethical topics such as constitutional law, tort law, assisted suicide, and new reproductive technologies, analysis of those topics is incidental to the task of analyzing how judges have invited, accepted, relied on, followed, critiqued, ignored, rejected, overridden, transformed, distorted, forced disclosure of, and otherwise responded to bioethics communications.

I am grateful to George Agich, who collaborated with me on my first exploration of bioethics expert testimony. I am also grateful to the bioethicists who generously shared their testimony and briefs with me, especially Judith Andre, the late John C. Fletcher, Thomas R. McCormick, John J. Paris, Clayton Thomason, and Robert M. Veatch. Thanks are also due to legal scholars David Jake Barnes, Edward Imwinkelried, Stephen Latham, Mary Anderlik Majumder, Lawrence J. Nelson, and philosopher Kenneth Kipnis, who participated in the Bridging Bioethics and Evidence Law Symposium at Seton Hall School of Law, and especially to my co-chair Ben A. Rich and Dean Kathleen Boozang, who generously supported that project. I owe an intellectual debt to John Monahan. Without Sarah Peters' efforts, the manuscript would not have been completed on time. Without the support of my husband, Keith Miller, it would not have been completed at all. This book is dedicated to him, with gratitude.

Bethany J. Spielman, JD, PhD

Footnotes

[1] Brown v. Board of Education, 347 U.S. 483 (1954).
[2] Wolf S. Law & Bioethics: From Values to Violence. J. Law Med. & Ethics 32: 293–301 at 300 (2004).

Contents

xi

Introduction

At the Interface of Bioethics and Law

1. Law's Receptivity to Bioethics

Legal disputes regarding issues ranging from stem cell research to implantable artificial hearts to assisted suicide and biobanking arise with increasing frequency. In the course of those disputes, bioethics resources appear in the forms of expert testimony, ethics committee determinations, the work of institutional review boards, and bioethicists' research materials. Even Supreme Court justices have cited some of these resources.[1] However, in a recent pretrial decision in a case to determine whether Merck, which manufactured the COX-2 inhibitor, Vioxx was liable for the death of a heart attack victim, a judge barred even the use of the words "ethics" and "morality." The judge commented, "That's not to say the trial isn't about what's wrong or right. That's what the law is about."[2] Instead of a welcome guest, bioethics testimony had become *persona non grata*.

Law and bioethics are inherently different social and communicative systems. Each constructs a social reality of its own,

From: *Bioethics in Law*
By: B. J. Spielman © Humana Press Inc., Totowa, NJ

communicates distinctive norms, and fills a different social function. Each has different goals, methods, and epistemologies. Each identifies and uses expertise, presumptions, values, and burdens of proof in distinctive ways,[3] yet they are deeply dependent on each other. One scholar has characterized them as "strange bedfellows."

The simultaneous separateness and mutual interdependence of law and bioethics raises important but difficult questions regarding their boundaries, relationships, and interface. Whenever nonlegal materials are borrowed for law's purposes—regardless of whether those materials are scientific, medical, social scientific, or bioethical—questions arise regarding how they interact with law, how closely law can rely on them, and how much openness or closure toward nonlegal material is desirable. Sociolegal systems theorists assert that, to function effectively in complex societies, law interacts with other systems in a variety of ways. These interactions are intricate formal and informal arrangements that link law with other systems. Law can, as a result, receive input from them; rely on them on an ongoing basis, even delegate some of its tasks to them; and thus evolve to meet its own needs and the needs of an increasingly complex society.[4]

However, although law is open to other systems, it cannot be completely open, any more than another system, such as bioethics, could be completely open to law.[5] If law were to borrow from or delegate to other systems unreflectively, it would abandon its character and identity as law, eroding core values of individual rights and procedural democracy.

Law's boundaries are said to be more porous if law is presented with cognitive rather than normative material. Bioethics' norms operate alongside law, but, if law functions well, bioethics' norms are not absorbed directly into law. Law selects, according to its own criteria, which external norms it will receive and how to use them, just as bioethics selects which legal norms it will receive and how to use them.

Some skepticism regarding drawing sharp distinctions between norms and other bioethics material is justified. However, when law confronts a system, such as bioethics, that has strongly normative features, some differentiation is critical, or law may be confused with religion or ethics. In the United States, limits are set on how open law can be toward religions. The much-contested First Amendment, premised on the fact of religious and moral pluralism, has historically protected the freedom of individuals to follow their own religious norms, in part by preventing law from endorsing any specific religion. There is no ethics corollary to the First Amendment's religion clauses. Despite early attempts to portray bioethics as "philosophy for the people"[6] and as an effort to "empower democracy,"[7] too much openness on the part of law to bioethics' normativity can be as troubling as too much openness to religion's normativity. Some observers of bioethics in the 1990s, drawing on First Amendment language, warned about an "establishment bioethics." One commentator, uneasy with the direction of the field, suggested this Constitutional amendment: "Congress shall make no law respecting an establishment of ethics, or prohibiting the free exercise thereof."..."If individuals are the arbiters of their own fate," he writes, "and are supposed to follow the dictates of conscience, then to have that conscience determined by a secular priesthood...is as offensive as having that conscience determined by a religious priesthood."[8] Today, the probability is higher that law will fail to set boundaries differentiating itself from a religiously based ethics, and endorse, or open itself too completely, to fundamentalist Christian norms,[9] but the issue of law's openness remains.

2. Bioethics' Eclecticism

To the extent that law differentiates between normative and other materials, judges will need to sort through the various components of bioethics communications. In *The Abuse of Casuistry*,

Jonsen and Toulmin describe moral reasoning as a composite of many loosely woven strands:

> Those who take a rhetorical view of moral reasoning…do not assume that moral reasoning relies for its force on single chains of unbreakable deductions which link present cases back to some common starting point. Rather (they believe), this strength comes from accumulating many parallel, complementary considerations, which have to do with the current circumstances of the human individuals and communities involved and lend strength to our conclusions, not like links to a chain but like strands to a rope or roots to a tree.[10]

The strands of bioethics are drawn from a variety of sources, methods, theories, and fields, combined in ways that are alternately multidisciplinary, interdisciplinary, and nondisciplinary. A recent book on medical ethics illustrates this eclecticism: it includes 13 "methods," which the editors compare with Wallace Stevens' 13 "ways of looking at a blackbird": philosophy, religion and theology, professional codes, legal methods, casuistry, history, qualitative methods, ethnographic methods, quantitative surveys, experimental methods, and economics and decision science.[11] In addition, some ethics literature subdivides ethics into three branches: normative ethics, which provides moral action-guides; descriptive ethics, which describes what people believe to be right and wrong; and metaethics, which investigates the nature of moral statements and reasoning.

An example of bioethics testimony illustrates how the products of multiple modes of inquiry, including moral norms, are used to create bioethics. This testimony is taken from *Biddison v. Facey Medical Group*,[12] a medical malpractice suit, in which a California man died while waiting for a replacement pacemaker. One of the issues about which a bioethicist testified was whether the physician defendant had a duty to disclose his own financial incentives to limit treatment options, including incentives to limit referral to an out-of-plan medical center at which Mr. Biddison might have received timely care. An attorney asks the ethics expert whether a patient has a right to be informed about the physician's financial incentives, and the expert answers, describing his reasoning:

Q. Let me talk about the personal financial considerations of the doctor or the group. If there's a conflict between that and the care being given to the patient, does the patient have the right to be informed about that?

A. That is the great ongoing debate. So far as I read in the literature and I read this up to date as of two weeks ago, there's no evidence in the literature and no evidence in the standard of care that physicians have …an obligation to inform patients of that conflict… [In] fee for service [medicine], the physician has an enormous conflict of interest in ordering up more tests; and the prime example we see of that is the Hickford Study done on imaging centers in Florida. They learned that 90% of all the imaging centers in Florida were physician owned and they were owned primarily by physicians that would refer patients to MRIs and those physicians referred those patients to the free standing facilities which they had financial interests at a rate of 40% higher than physicians in academic medical centers.[13]

A. …Do you want us to set up a system whereby we inform the physicians and the physician looks the patients in the eye and say, "You of course realize that when I tell you that I do not recommend this surgery, that I have a potential conflict of interest at stake here, so you should be very suspicious and you might want to get a second opinion and you might want to go outside the system to get the procedure." That's no way to practice medicine.[14]

The expert neatly summarized the strands of his eclectic testimony as follows: "So do I think they should inform the patients? I think A, it's not done. B, the literature search at least two weeks ago indicates that it's never done and C, I think it would have a terribly detrimental effect on the patient–physician relationship."[15]

The "A," "B," and "C" strands of reasoning help in reaching the conclusion that the physician had no moral obligation to disclose his financial incentives to Biddison. They were produced with moral norms, but not only with moral norms. The "A" strand, resulting in the empirical claim "It's not done," is based, according to the expert's previous testimony, on the expert's observations of physicians and physician–patient interactions. The "B" strand (a literature search indicates it's never done) also results in an empirical claim, but is based on literature regarding health care delivery. The "B" strand also generalizes from the group studied in the research (especially the "Hickford [sic] study") to other physicians.[16] The "C" strand ("it would have a terribly detrimental effect on the patient–physician relationship") is both a prediction regarding consequences for patient–physician relationships and a judgment based on the moral norm, "patients should trust physicians."

3. The Approach of This Book

Like the transcript excerpt above, the examples that will be analyzed in this book are bioethics communications. They have been used in, or offered for use in, proceedings before judges during the last decade. How receptive judges, in particular, are to bioethics' norms—as well as to the other features of bioethics—is critical because of the importance of judicial opinions to our legal system. Judges have been the "primary sources" of law in our system of government for 200 years. The law is what judges say it is.[17] Further, judicial decisions have an immediate force that can profoundly affect the lives of individuals and family members, access to potentially life-saving medical treatment, and freedom from institutionalization, as well as the transfer of large monetary sums.

The organization of this book is unlike that of other volumes that treat bioethics and law. It touches on traditional bioethics topics, such as assisted suicide, human subjects research, and reproductive technologies. It will also touch on traditional legal subjects, such as constitutional, criminal, and tort law. However, legal doctrine and bioethics argument are incidental to the main task. The basic unit of analysis is an interaction between a particular type of bioethics communication and the legal system. Rather than organize the book according to bioethics topics or conventional legal categories, therefore, the book is organized according to these communications.

This volume includes six types of bioethics communications.[18] The first two, expert bioethics testimony and *amicus curiae* briefs, are distinctive among bioethics communications in that they are means by which law meets bioethics. They are introduced in Chapter 1, which explores the fundamental question, when do judges find bioethics helpful? Chapters 2 through 5 examine bioethics communications that, unlike expert testimony and *amicus curiae* briefs, are not developed to interact with the legal system but to fill bioethics' other social functions. Chapter 2 examines health care ethics committee determinations; Chapter 3, the determinations of institutional review boards; Chapter 4, federal bioethics commission reports; and Chapter 5, subpoenaed bioethics research. Chapters 6 through 8 return to bioethics expert testimony, exploring, in depth, the evidentiary requirement that testimony be reliable. Chapter 6 explores "general acceptance" as an indicator of the reliability of expert testimony. Chapter 7 examines peer-reviewed publication as a criterion of its reliability. Chapter 8 assesses experience as an indicator of its reliability. Taken as a whole, these eight chapters describe how judges may invite, accept, rely on, follow, critique, ignore, override, transform, distort, force disclosure of, and otherwise interact with bioethics communications. A final comment assesses the future of bioethics in law.

This exploration of bioethics in law is intended to be a resource for attorneys, judges, and law clerks who have occasion to evaluate and use expert bioethics testimony, bioethics commission reports, research materials, ethics committee determinations, and other bioethics communications. It is also intended to help bioethicists and health care professionals who provide ethics resources for use in legal proceedings. Finally, it is for scholars of evidence law and legal studies, and for students of the humanities and others committed to observing and analyzing bioethics in law.

Endnotes

[1]Cruzan, by her parents and co-guardians, Cruzan et ux. v. Director, Missouri Department of Health, et al., 497 U.S. 261 (1990). Irving Rust, et al., Petitioners v. Sullivan, Louis W., Secretary of Health and Human Services; New York, et al., Petitioners v. Louis W. Sullivan, Secretary of Health and Human Services, 500 U.S. 173 (1991). Bowen, Secretary of Health and Human Services v. American Hospital Association, et al., 476 U.S. 610 (1986). Chief Justice Rehnquist cited the Brief for Bioethics Professors in Washington, et al., Petitioners v. Harold Glucksberg, et al., 521 U.S. 702, 733 n. 23 (1997).

[2]Vioxx: New Jersey Judge Bars Testimony on 'Ethics' 'Morality.' American Health Line. September 9, 2005. Available at LEXIS NEXIS.

[3]Legal reasoning is more constrained by dictates of history, precedent, and institutional concerns than is bioethics. *See*: Rao N. A backdoor to policy making: The use of philosophers by the Supreme Court. U Chicago L Rev 1998;65:1371–1401; Collier CW. The use and abuse of humanistic theory in law: Reexamining the assumptions of interdisciplinary legal scholarship. Duke L J 1991;41:191–272; Fallon RH Jr. Non-legal theory in judicial decisionmaking. Harv J L & Pub Pol'y 1994;17:87–99; Fiss OM.

The death of law? Cornell L Rev 1986;72:1–16; Posner RA. The problems of jurisprudence. Cambridge, MA: Harvard University, 1990; Posner RA. Conceptions of legal "theory": A response to Ronald Dworkin. Ariz St L J 1997;29:377–388. *See also*: Collier CW. The use and abuse of humanistic theory in law: Reexamining the assumptions of interdisciplinary legal scholarship. Duke L J 1991;41:191–272.

[4]Luhmann N. Law as a Social System, Fatima Kastner, RN; David Schiff, editor, Klaus A. Ziegert, trans. Oxford Socio-Legal Studies. New York: Oxford University Press, 2004; Luhmann N. Law and social theory: Law as a social system. NW U L Rev Fall 1988/Winter 1989;83:136–150;Teubner G. Communicative rationalities in law, morality, and politics. Cardozo L Rev 1996;17:901–918; Teubner G. The global interplay of legal and social systems. Am J Comp L Winter 1997;45:149–169; Teubner G, Bremen F. Introduction to autopoietic law. In: Teubner G, ed. Autopoietic law: A new approach to law and society. Berlin; New York: Walter de Gruyter & Co., 1987:1–11 at 4; Teubner G. How the law thinks: Towards a constructivist epistemology of law. Law & Soc'y Rev 1989;23:727–757 at 739; Sand IJ. Changing forms of governance and the role of law-Society and its law, http://www.arena.uio.no/publications/working-papers2000/ papers/wp00_14.htm (visited January 25, 2006).

[5]The fifth edition of Tom L Beauchamp's and James F. Childress' classic Principles of Biomedical Ethics illustrates. It incorporated only two legal cases that had not appeared 7 years before in the 4th edition. Neither Washington v. Glucksberg nor Vacco v. Quill, two obviously important legal decisions, were included. These cases were not necessary to, and perhaps would have conflicted with, the authors' purpose, which was to develop an account of bioethics. Beauchamp TL, Childress JF. Principles of biomedical ethics. 5th ed. New York, NY: Oxford University Press, 2001.

[6]Jonsen AR, Jameton A. Medical ethics, history of: The Americas: Academic bioethics. In: Reich WT, ed. Encyclopedia of Bioethics: Revised Edition, vol. 3. New York: Simon & Schuster Macmillan, 1995:1626–1632.

[7]Bloche G. Medical ethics in the courts. In Danis M, Clancy C, and Churchill L, eds. Ethical Dimensions of Health Policy. New York: Oxford University Press, 2002:133–156 at 149.

[8]Scofield G. Post modernist government, MCW discussion list, Dec. 3 and 4, 1996.

[9]First Amendment claims were raised explicitly in Rideout v. Hershey Medical Center, 30 Pa. D. & C. 4th 57 (1995), and In the Matter of Baby "K," 832 F. Supp. 1022 (1993), which are discussed in Chapters 2, 6, and 7.

[10]Jonsen AR, Toulmin SE. The abuse of casuistry: A history of moral reasoning. Berkeley: University of California Press, 1988:293–294.

[11]Methods in medical ethics, Sugarman J, Sulmasy DP, eds. Washington, D.C.: Georgetown University Press, 2001 at 286.

[12]Dorothy Biddison, James C. Biddison, etc., Plaintiffs, v. Facey Medical Group, Khosro Sadeghani, M.D., Northridge Hospital, Niles Chapman, M.D., Take Care Health Plan, and Does 1-100, Inclusive, Defendants, Case No. PC016239X. Deposition of John J. Paris, S.J., in Irvine, Cal. (June 17, 1998).

[13]Dorothy Biddison, James C. Biddison, etc., Plaintiffs, v. Facey Medical Group, Khosro Sadeghani, M.D., Northridge Hospital, Niles Chapman, M.D., Take Care Health Plan, and Does 1-100, Inclusive, Defendants, Case No. PC016239X, deposition of John J. Paris, S.J., on June 17, 1998, p. 79 line 7 to p. 80 line 4.

[14]Dorothy Biddison, James C. Biddison, etc., Plaintiffs, v. FaceyMedical Group, Khosro Sadeghani, M.D., Northridge Hospital, Niles Chapman, M.D., Take Care Health Plan, and Does 1-100, Inclusive, Defendants, Case No. PC016239X, deposition of John J. Paris on June 17, 1998, p. 81 line 19 to p. 82 line 2.

[15]Dorothy Biddison, James C. Biddison, etc., Plaintiffs, v. Facey Medical Group, Khosro Sadeghani, M.D., Northridge Hospital, Niles Chapman, M.D., Take Care Health Plan, and Does 1-100, Inclusive, Defendants, Case No. PC016239X, deposition of John J. Paris on June 17, 1998, at p. 82 line 3 to page 82 line 7.

[16]No "Hickford" study is reported in the literature. The expert may be referring to Hillman BJ, Joseph CA, Mabry MR, et al. Frequency and costs of diagnostic imaging in office practice—A comparison of self-referring and radiologist-referring physicians. N Eng J Med 1990;323:1604–1608.

[17]Marbury v. Madison, 5 U.S. (1 Cranch) 137 (1803).

[18]Tod Chambers observes that the ethics case story is bioethics' key genre. That genre is not treated here as a bioethics communication because the bioethics case story *per se* does not interact with law. Chambers T. What to expect from an ethics case (and what it expects from you). In: Nelson HL, ed. Stories and their limits: Narrative approaches to bioethics. Routledge: New York and London, 1997:171–184.

1 How Does Bioethics Help Judicial Reasoning?

Each of the bioethics resources to be analyzed in Chapters 3 through 6 can interact with law, but to do so it needs a vehicle by which it can reach the legal system. This chapter provides an overview of the helpfulness of two important vehicles: bioethics expert testimony and *amicus curiae*, or friend of the court, briefs.

Law looks to nonlegal resources to obtain help. Therefore, it is not surprising that rules governing both expert testimony and *amicus curiae* briefs ("briefs") allude to, or explicitly require, helpfulness. The very first condition that Federal Rule of Evidence 702, which governs expert testimony in federal courts, establishes for admission of such testimony is that: "[S]cientific, technical, or other specialized knowledge...will assist the trier of fact to understand the evidence or to determine a fact in issue."[1] Likewise, Supreme Court Rule 37.1, governing the filing of briefs, specifies: "An *amicus curia* brief that brings to the attention of the Court relevant matter not already brought to its attention by the parties may be of considerable help to the Court. An *amicus curiae* brief that does not serve this purpose burdens the Court, and its filing is not favored."[2]

From: *Bioethics in Law*
By: B. J. Spielman © Humana Press Inc., Totowa, NJ

To understand how expert testimony or a brief might meet these expectations for helpfulness, consider what judges do. They engage in practical reasoning to accomplish two fundamental tasks: determining facts and making law. Distinguishing between the former, adjudicative function, and the latter, legislative function, is important for understanding in what ways bioethics communications may help. Kenneth Culp Davis observed:

> [When a court] finds facts concerning immediate parties— what the parties did, what the circumstances were—[it] is performing an adjudicative function, and the facts may conveniently be called adjudicative facts. When [a court] wrestles with a question of law or policy, it is acting legislatively, and the facts which inform its legislative judgment may conveniently be denominated legislative facts."[3]

Davis' distinction between legislative facts and adjudicative facts was incorporated into the Federal Rules of Evidence.[4] Federal Rule of Evidence 201 indicates that the evidentiary rules apply to adjudicative facts (the "who, what, when, where, why, how, and with what motive or intent" of the case), but not to legislative facts.[5] Judges, therefore, have greater latitude to accept nonlegal material if legislative facts are being considered than if adjudicative facts are being determined. Although Davis wrote before the modern era of bioethics, and some today reject that the notion that a "fact" could include material that even hints at normativity,[6] bioethics testimony may, nonetheless, occasionally help determine adjudicative facts, as in the patent infringement case presented next.

1. How Can Bioethics Testimony Help?

1.1. Testimony Helps Determine Adjudicative Facts

Zenith Lab. v. *Bristol-Myers Squibb Co.*[7] illustrates how bioethics testimony has been used to help determine an adjudicative fact regarding research. *Zenith Lab.* v. *Bristol-Myers Squibb Co.,* which was decided in 1992 by the US District Court of the District of New Jersey, involved a complaint under US patent laws.[8] The question before Judge Wolin was whether the antibiotic, Cefadroxil DC, for which Zenith Laboratories had obtained Food and Drug Administration (FDA) approval, infringed a patent held by Bristol-Myers Squibb. Zenith sought a declaration that it did not infringe Bristol-Myers Squibb's patent for Bouzard monohydrate. Bristol-Myers Squibb claimed that Cefadroxil DC did infringe the patent—not in its manufactured, preingested state—but after ingestion. Bristol-Myers Squibb claimed the in vivo conversion rendered Cefadroxil DC an infringing compound. The primary factual issue at trial was whether Cefadroxil DC converted in vivo to the Bouzard monohydrate.

The evidence consisted, for the most part, of testimony by experts who had conducted experiments either to determine whether Cefadroxil DC converts in vivo to the Bouzard monohydrate or to support or rebut assumptions, methods, or conclusions of those experts. Several experts testifying for Bristol-Myers Squibb had conducted in vitro experiments that they believed were simulations of in vivo conditions adequate to demonstrate conversion of the compound. None had actually conducted in vivo experiments. Robert Levine, a medical ethicist, testified as an expert witness on behalf of Bristol-Myers Squibb that this would have been impossible; an institutional review board (IRB) would not have approved in vivo experimentation merely to prove patent infringement. He also testified that he did not think that an in vivo experiment could be designed to test the conversion hypothesis that would be any more probative of the issue than were the in vitro studies that were intended to simulate in vivo conditions and that had already been conducted.

Although Zenith had presented several defenses, the court ultimately entered a judgment in Bristol-Myers Squibb's favor. Levine's testimony clarified for the court the ethical constraints on in vivo experimentation under which Bristol-Myers Squibb had been operating. In effect, Levine speculated that any IRB would have rejected as unethical the kind of research Zenith Labs suggested was necessary. Without Levine's testimony, the court might have had an unrealistic expectation that Bristol-Myers Squibb could have performed in vivo experiments with Cefadroxil DC, instead of simulations to demonstrate conversion to the Bouzard monohydrate.[9] The ethics testimony provided evidence of the fact that conducting in vivo research to demonstrate conversion would have been impossible under current ethical constraints. Judge Wolin needed this information to understand the ethical limits of human experimentation, and ultimately to decide that Bristol-Myers Squibb had met its burden of proof to establish patent infringement.

1.2. Testimony Provides Legislative Facts

Legislative facts are more general than adjudicative facts. They transcend the particular case. *Planned Parenthood Fed'n of Am. v. Ashcroft*[10] explicitly addresses this difference. The constitutionality of the Partial-Birth Abortion Act of 2003 was before the US District Court for the Northern District of California. The Act prohibited physicians from performing what the Act called "partial-birth abortion." Plaintiffs, who sought an injunction permanently enjoining the enforcement of the Act, contended that the Act placed an undue burden on a woman's right to choose; was impermissibly vague; violated a woman's due process right to bodily integrity; and violated the Fifth Amendment. We focus here on the claim that the Act failed to provide an exception for the health of the mother, violating a woman's Fifth Amendment due process rights as set forth by the Supreme Court in *Planned Parenthood v. Casey*,[11] and *Stenberg v. Carhart*.[12]

Judge Hamilton's reasoning regarding *Stenberg v. Carhart's* "medical necessity" exception for the health of the mother was aided by medical ethics testimony. Congress had made a set of ethics-related findings in support of the Act. It had found, for example, that there was a medical consensus that dilation and evacuation (D&E) was unethical; that informed consent for D&E was impossible; that the Act preserved the integrity of the medical profession, and that the Act was "not required to contain a health exception...because a partial birth abortion is never necessary to preserve the health of the woman." Whether the findings were treated as adjudicative facts or legislative facts could affect the case outcome. If these Congressional findings were findings of adjudicative fact, then the court would have been expected to afford them substantial deference; if they were legislative facts, then the court would not be expected to give the findings deference.[13]

The government wanted to characterize *Stenberg* v. *Carhart's* health exception as an adjudicative fact, one that concerned only the immediate parties to the dispute. That approach, however, would have created the prospect of different jurisdictions with different constitutional practices, despite the fact that the empirical issue (whether D&E is ever necessary for the woman's health) was identical in each of them.[14] *Stenberg* v. *Carhart* had rejected that approach.

The district court recognized that, to produce the nationally uniform approach that *Stenberg* v. *Carhart* required, it would have to assess the constitutionality of the matter of the necessity of a health exception at the level of legislative fact rather than at the level of adjudicative fact. As a result, the court did not defer to the above Congressional findings. Instead, it followed *Stenberg* v. *Carhart's* holding that the existence of a division of medical opinion supports the need for a health exception. The court used its own evidence regarding the medical necessity exception, in addition to the evidence that Congress had used.[15]

That evidence, including medical ethics testimony, demonstrated that there was no ethical consensus regarding the appropriateness of D&E, and that a ban would not promote, but, in fact, could erode, the integrity of the medical profession. One expert testified that even his own colleagues did not share his ethical views regarding the procedure. Other experts discussed the "extraordinarily negative impact that the Act would have and has had on their relationship with their patients and on their ability to provide the care that they deem to be in their patient's best interests."[16]

The medical ethics testimony contradicted Congressional findings and, thus, helped demonstrate a lack of ethical consensus among physicians. Under *Stenberg* v. *Carhart*, that lack of medical consensus necessitated a health exception. As a result, the court held that the Partial-Birth Abortion Act's lack of an exception violated the Fifth Amendment, and permanently enjoined enforcement of the Act.

The ethics testimony by no means stood alone as a reason for the case outcome in either *Planned Parenthood Fed'n of Am.* v. *Ashcroft* or *Zenith Lab.* v. *Bristol-Myers Squibb Co.* In *Planned Parenthood Fed'n of Am.* v. *Ashcroft*, the district court relied not only on bioethics testimony, but also on a variety of other testimony, and found not only that the Act violated the Fifth Amendment, but also that the Act would place an undue burden on a woman's right to choose, and was impermissibly vague. In *Zenith Lab.* v. *Bristol-Myers Squibb Co.*, the court relied not only on the medical ethics expert, who could explain the ethical constraints preventing Bristol-Myers Squibb from conducting in vivo experiments, but also on the testimony of numerous scientific experts who explained the simulated in vivo experiments that Bristol-Myers Squibb had conducted, as well as on patent law precedent. However, the fact that ethics testimony is not pivotal in a case does not mean it is not helpful or even important to judicial reasoning.

1.3. Can Testimony Provide "Normative Facts"?

At this point, the reader may wonder about a third category of potential uses for bioethics testimony: "normative fact." The term, although linguistically parallel to Davis' "adjudicative fact" and "legislative fact," is not one that Davis used, and there is a legitimate debate regarding whether the notion of "fact" as

Davis understood it is elastic enough to include products of normative judgment. Further, although the term normative fact is used on occasion by legal scholars and philosophers of law, it tends to be defined differently by each commentator, and often to be assigned narrower meanings than would be helpful in this book. I will, nonetheless, suggest that the term normative fact be used when judges treat bioethics adjudicative or legislative facts as normative law.

Bioethics communications occasionally function as normative facts in this sense. However, bioethics testimony does so only under unusual circumstances. We saw an example of such testimony previously, in the Zenith Lab. example. To make a determination regarding burdens of proof, Judge Wolin had to consider the legal character of IRBs, the potential medical risks, and the law on burdens of proof, as well as the expert's normative judgment regarding what IRBs should do. He could be informed by the expert's knowledge and normative judgment, but could not substitute the expert's normative ethical judgment for his own, or for the law on burdens of proof or patent infringement. Testimony that could serve to displace, distract from, or be confused with, these legal judgments and norms will not be received well, as discussed in the next section.

2. When Is Bioethics Testimony Unhelpful?

2.1. Testimony Is Irrelevant

Bioethics communications—even those developed specifically for litigation purposes—are not always helpful. Expert testimony that is not relevant to the task at hand, contradicts law, or directly challenges legal boundaries can fail to help.

It might seem initially that bioethics could help whenever certain kinds of cases are adjudicated, for example, end-of-life medical treatment cases, or "cutting edge" bioethics cases.[17]

However, the rules of evidence suggest otherwise. In *Daubert* v. *Merrell Dow Pharm.*, the Supreme Court repeated what is obvious, based on the rules of evidence: that relevance is a condition of helpfulness. It wrote: "Expert testimony which does not relate to any issue in the case is not relevant and, ergo, non-helpful."[18] The Federal Rules define relevance as: "evidence having any tendency to make the existence of any fact that is of consequence to the determination of the action more probable or less probable than it would be without the evidence."[19]

Even in end-of-life cases (regarding which, bioethics has much to say), bioethics testimony may be irrelevant to law's tasks. An example is bioethics testimony in the much-publicized Terri Schiavo case, which was actually a series of legal actions stretching over a decade. *Schindler* v. *Schiavo* (*In re Schiavo*) was a 2001 appeal from an order authorizing discontinuing nutrition and hydration from the 38-year-old Florida woman in a persistent vegetative state.[20] Florida statutory and case law governed termination of life-prolonging procedures. Schiavo's husband and guardian had invoked the trial court's jurisdiction to allow the trial judge to serve as the surrogate decision maker.[21] The relevant legal standard was the substituted judgment standard for surrogate decision making for patients incapable of making decisions regarding their own medical care. The trial court's task, under Florida law, had been to determine what the preferences of Ms. Schiavo would have been under her current circumstances.

A bioethics expert gave testimony that was not directly relevant to the substituted judgment standard. She offered the results of a survey that asked about respondents' end-of-life preferences. There was no evidence that Ms. Schiavo had participated in the survey. Instead of providing material to help answer the question, "What would Ms. Schiavo have wanted?" the bioethicist's presentation answered the question, "What do other people say they want?" On appeal, the parents of Ms. Schiavo complained of the bioethics testimony. The Schindlers claimed that the trial court should not have heard the survey evidence from the bioethicist. The appellate judge, agreeing with the parents, summarized his view of the testimony:

> We have considerable doubt that [the bioethics expert's] testimony provided much in the way of relevant evidence. She testified about some social science surveys. Apparently most people, even those who favor initial life-supporting medical treatment, indicate that they would not wish this treatment to continue indefinitely once their medical condition presented no reasonable basis for a cure. There is some risk that a trial judge could rely upon this type of survey evidence to make a "best interests" decision for the ward. In this case, however, we are convinced that the trial judge did not give undue weight to this evidence and that the court made a proper surrogate decision rather than a best interests decision.[21]

The appellate judge thought that relying on the bioethicist's testimony would have been a mistake. Ethics testimony regarding what other people would have wanted if they were in Ms. Schiavo's situation was beside the point, because the jurisdiction required that a substituted judgment be made and did not permit imputation of beliefs and values to others. The appellate judge was convinced, however, that the trial judge had not given much weight to the best interests-oriented bioethics testimony and that he had relied, as he should have, on Ms. Schiavo's own oral statements to her friends and family.

Some bioethics testimony is irrelevant not because it addresses the wrong question, as in *Schindler* v. *Schiavo* (*In re Schiavo*), but because it is in some other way a mismatch for the practical needs of the court. In another recent Florida case, *Hall* v. *Anwar*, the bioethics testimony had erroneously been admitted into evidence; state law required that only medical testimony could state the standard of care. In addition to this problem, the bioethics testimony presented another difficulty that is of greater interest here: it was too abstract to be relevant. *Hall* v. *Anwar* was an appeal of the admission of bioethics testimony in this case.[22] The parents of a severely brain-damaged infant had claimed that health care providers were negligent when they stopped an effort to resuscitate their prematurely born infant after 11 minutes, declared the infant dead, and, discovering that the infant was not dead 15 minutes after that, started resuscitation efforts again. Concluding that admission of a bioethicist's testimony had been erroneous (because the expert was not a physician, as Florida law required), the judge also determined that the admission error was harmless because the lower court judge had made no use of the testimony. This testimony, although perhaps philosophically important, was too tenuously related to the legal case to be considered relevant. Referring to the bioethicist, Judge Altenbern stated:

> His testimony was often very abstract, describing such things as the 'metaphysical' and 'epistemological' issues associated with the 'post-Kantian world' and its view that 'perception is the real.'…It is not surprising that all of the lawyers essentially ignored this testimony during closing arguments.[23]

The relation between this testimony and any matter that could be proven in the case was too ill defined to be used by attorneys, or to be admissible in the eyes of Judge Altenbern. In addition to the problems of expert qualification and of relevance, it is also worth noting how Judge Alternbern struggled over the expert's status as a Jesuit priest. He noted the fact that the expert was a priest, but also that the expert had not testified in religious attire, emphasized his position as a priest, or concentrated on religious doctrine.[24] Had the expert's presentation been overtly religious, presenting the norms of a particular religion, Judge Altenbern might not have found its admission harmless. Given the expert's nonreligious presentation, however, it created only the problems of relevance and proper qualification of the expert, and was a harmless error.

2.2. Testimony Contradicts Law
or Challenges Legal Boundaries

One of the most difficult questions in the area of bioethics in law is when bioethics norms can be used in law. At least this much is clear: law will be unreceptive to bioethics testimony that conflicts with legal standards because it presents the risk that legal decisions will be made on improper grounds.[25] The potential for direct conflict between bioethics norms and legal norms is illustrated in an excerpt from a bioethics expert's deposition in the managed care case introduced in the Introduction, *Biddison* v. *Facey*.[26] A 42-year-old accountant had received a pacemaker at Yale University. The lifespan of the pacemaker was 8 years. The man had had the pacemaker replaced twice before at Yale University, where it was originally implanted. This time the man went to his medical group, selected through his health insurance, and requested that it be replaced again at Yale. Because of a miscommunication between the primary care physician in the medical group and the cardiothoracic surgeon, the pacemaker was never ordered. The accountant died while waiting for surgery. In a malpractice case brought after the man's death, a bioethics expert testified for the managed care medical group. The ethicist responded to a question from the attorney regarding the physician's obligation to disclose to patients his incentives to limit treatment by not referring to providers outside the plan, such as Yale University. Recall that the bioethicist responded:

A.   …So do I think they should inform the patients? I think A, it's not done. B, the literature search at least two weeks ago indicates that it's never done and C, I think it would have a terribly detrimental effect on the patient–physician relationship.

By Mr. Heimberg:

Q. Are you aware that the position you just cited is in
    direct conflict with the law of California?

A. I'm not aware of that.

Q. ...to the extent that it's in conflict with the laws of
    California, prior to the time of trial and testifying, are
    you going to do research on that?

A. No. I'm not going to testify as to what the law of the
    State of California is, and if the judge thinks that
    what I'm testifying to is contradictory to law, he will
    instruct me.

Mr. Crandall: Although the witness has been provided with
a copy of Section 1367.

The Witness: I was just provided with it. I'll read it, of
course. But remember as the Supreme Judicial Court of
Massachusetts said, 'The law frequently lags behind developments in medicine and looks to philosophy, theology and
public policy for guidance and insight as to how it should
develop.'

Mr. Heimberg: Can you say that one again?

The Witness: Yes. You have to be cognizant of the insight of
the Supreme Judicial Court of Massachusetts gave in the
Satewoods' (ph.) opinion when it says that the law frequently lags behind developments in science, technology
and medicine and the law looks to philosophy for guidance
and insight as to how the law could more appropriately be
formulated.[27]

Notice that attorney Heimberg implies that bioethics
testimony contradicting legal norms cannot help law, and
that the bioethics expert implies that such testimony can. The

disagreement between the attorney and the bioethicist reflects a difference of opinion regarding the conditions under which law, which has normative components, should be open to normative components in bioethics.

To support his belief that bioethics could help, the bioethicist invoked language from a Massachusetts opinion that had appreciatively noted the contribution of ethics.[28] He invoked *Superintendent of Belchertown State School* v. *Saikewizc*, a 22-year-old case that had presented novel issues to a Massachusetts court. Ethics had been used in that case to create a framework for decision making in an area in which there was no substantive governing law (a legislative purpose). In *Biddison*, the opposing attorney claimed that a legal standard requiring disclosure already governed. Yet the ethics expert was suggesting that another standard—one the attorney thought conflicted with the legal standard—be used. A judge might modify or extend governing law under certain circumstances, and even identify bioethics norms as one reason for doing so, if the bioethics material did not contravene legal norms. If bioethics testimony directly contradicts law, a judge would be far more likely to reject or ignore the testimony than to ignore the law.

Bioethics norms cannot only directly contradict law's norms, as attorney Heimberg thought the testimony in *Biddison* did. Testimony can also create confusion between the law's norms and bioethical norms, as well as subvert the processes by which law allocates normative decision making. In some instances, judges reject the testimony on the basis that it may confuse or mislead a jury. Federal Rule of Evidence 403 permits relevant evidence to be excluded "if its probative value is substantially outweighed by the danger of unfair prejudice, confusion of the issues, or misleading the jury, or by considerations of undue delay, waste of time, or needless presentation of cumulative evidence."[29]

This rule has particular salience when legal language has been assimilated by bioethics, or vice versa. When that occurs, the original meaning is often transformed—some would say, distorted—to meet the needs of the system into which it is imported. The language may become what philosopher Judith Andre calls "bioethics as pidgin."[30] Sociolegal theorists remind us that language cannot be straightforwardly translated from one system to another. Such "borrowed" language was the subject of a Rule 403 objection in *In re: Rezulin Product Liability Litigation*, a products liability case for silent liver damage.[31] Judge Kaplan, a US district court judge in New York, considered the objection. Ethics experts intended to testify that Warner-Lambert, which developed the drug, acted in an unethical manner, especially with respect to clinical data and the conduct of clinical trials.

Judge Kaplan thought the ethics testimony would likely cause confusion by proposing alternative and improper grounds for a decision:

> … it would be likely unfairly to prejudice and confuse the trier by introducing the 'experts' opinions and rhetoric concerning ethics as alternative and improper grounds for decision on bases other than the pertinent legal standards. Accordingly, plaintiffs are precluded from offering any testimony, including that cited in the margin, concerning ethical standards and the application of ethical standards to the alleged conduct of the defendants and others.[32]

What was the "rhetoric concerning ethics" that Judge Kaplan thought might cause confusion? It was a "pidgin"—language that had, according to the judge, been borrowed from law by bioethics, and then been presented to law as "ethics."

> Dr. Furberg's opinions on the 'three basic rights' of patients are at best thinly-disguised legal or quasi-legal principles. This is particularly evident in the case of the so-called 'principle of self-determination,' which is nothing but a formulation of the doctrine of informed consent. ccordingly, Dr. Furberg's testimony on the 'basic rights of

patients' communicates a legal standard and so would encroach on the court's prerogative to instruct on the law.[33]

Judge Kaplan described as "semantic sleight-of-hand" the contention that language regarding "basic rights," the "duty to warn," and the "principle of self-determination" was not legal language.[34] Underneath the rhetoric, the language was legal language, in his view.

Of importance for purposes of this book is the boundary between legal norms and bioethics norms that Rule 403 was used to maintain in this case, which Judge Kaplan thought the plaintiffs had tried to subvert by intentionally misdescribing the testimony. Judge Kaplan was convinced that the norms offered in testimony were actually legal norms disguised as ethics, and, thus, an improper subject for expert testimony. The expert offering such testimony would usurp the judges' role as norm articulator.[35] Had the testimony used language that was more clearly ethical, Judge Kaplan would have faced a slightly different problem, although confusion would still have been a risk.

Expert ethics testimony can also be unhelpful when it uses traditional legal reasoning; that is the task of the parties' attorneys.[36] For example, in an Eighth Amendment prison health care case from Tennessee, *Bowman* v. *Corr. Corp. of Am.*,[37] a bioethics expert's report was inadmissible because the expert failed to distinguish adequately between his own role of ethics expert and the role of the attorney who retained him. The district court judge wrote that "[the] expert report read like a lawyer's brief."[38] Fearing that if the testimony were admitted, there might be grounds for reversal in the event of an appeal, the judge ruled the testimony inadmissible. In *Bowman*, as in *Biddison* and *In re Rezulin*, bioethics testimony was unhelpful because those offering it expected that a normative presentation from outside law could override or provide a substitute for law's own system of norms.

3. How Can Bioethics *Amicus Curiae* Briefs Help?

3.1. Briefs Provide Relevant Material Not Included in the Party Briefs

Amicus curiae briefs, a second type of bioethics communication developed specifically for legal purposes, can also
provide assistance to courts.[39] They may present an argument
or information, context, or an analytical approach not found in
the briefs of the parties. Two bioethics briefs illustrate this
point: the Brief for Bioethics Professors *Amicus Curiae*
Supporting Petitioners in *Vacco* v. *Quill* and *Washington* v.
Glucksberg,[40] the assisted suicide cases (Brief for Bioethics
Professors); and the Brief of Scholars in Medical Ethics as
Amici Curiae *In re AMB*,[41] involving medical treatment of a
severely disabled infant.

The Brief for Bioethics Professors was written by a group
of 50 professors who teach medical ethics. The brief argued that
the Ninth and Second Circuit Courts of Appeals were wrong in
failing to recognize that there are important differences between
the right to refuse medical treatment and the right to physician-
assisted suicide; and that the Court could refuse to recognize a
constitutional right to physician-assisted suicide and still recognize both constitutional and common law rights to refuse treatment. The brief also argued that it was a mistake for the Ninth
Circuit Court of Appeals to derive a right to assisted suicide from
cases directed at enabling a woman to choose abortion; that there
is no basis in logic or constitutional law to limit the right to
assisted suicide to terminally ill patients; and that the Court
could refuse to recognize a constitutional right to physician-
assisted suicide and still recognize a woman's constitutional
right to decide whether or not to terminate a pregnancy.

Chief Justice Rehnquist cited the Brief for Bioethics
Professors in his opinion for the court in *Washington* v. *Glucksberg*,

which held that Washington's prohibition against "causing" or "aiding" a suicide does not violate the Due Process Clause. In his discussion of the State's interest, he speculated that "the State may fear that permitting assisted suicide will start it down the path to voluntary and perhaps even involuntary euthanasia."[42] In its discussion of the slippery-slope argument, The Ninth Circuit Court had identified the belief that "terminal illness" could not be defined as a false premise, concluding that "While defining the term 'terminally ill' is not free from difficulty, the experience of the states has proved that the class of the terminally ill is neither indefinable nor undefined..."[43] As if to rebut this critique, the Brief for Bioethics Professors asserted that terminal illness is an "undefined and undefinable term."[44] If this premise were accepted as true, there could be no principled basis for confining a right to terminally ill patients. Justice Rehnquist accepted the premise, using the brief's assertion regarding terminal illness in reasoning that a right to physician-assisted suicide could not be contained.

The party briefs had not made the peculiar claim that terminal illness is not susceptible to definition. The Brief for Bioethics Professors' brief did, and it was this claim among all of the materials in their brief that Justice Rehnquist found helpful in his reasoning.[45]

3.2. Briefs Serve as Dialog Partners

Judges occasionally invite bioethicists to write briefs. On these occasions, a competently written bioethics brief, especially one that responds directly to questions put to it by the court, will help the court by serving, minimally, as an implicit dialog partner. A Michigan appellate judge's request prompted an example of such a brief. Judge Whitbeck requested a brief from medical ethicists for *In re AMB*,[46] which involved the following facts. In 1999, an infant had been born 5 weeks prematurely, with heart and brain abnormalities. The infant's 17-year-old mother was allegedly incompetent. The infant's putative father, who was also the father of the infant's mother, had been incarcerated; family agencies had become involved in the case. The hospital faced the question of whether to discontinue life-sustaining treatment. Within a few days of the infants' birth, the Family Independence Agency obtained an order authorizing hospital staff to "take the child off life support equipment and medication provided that 'Comfort Care' is provided." Despite a warning that the order did not take effect for 7 days, hospital staff removed the infant from life support the next day, and the infant died. The attorney appointed to represent the infant in the protective proceeding that originally brought her situation before the family court appealed.

When the case was on remand from the Michigan Supreme Court, Judge Whitbeck faced, and asked the medical ethicists to address, several questions. These questions concerned the standards and procedures for determining parental competence for making medical decisions for children, surrogate decision making, and how the decision to withdraw life-sustaining treatment should be made. The bioethics brief responded directly to each of the questions the court had posed.

Both the brief and the opinion Judge Whitbeck wrote for the appeals court followed mainstream thinking in law and bioethics regarding standards for surrogate decision making. However, in other areas, Judge Whitbeck's conclusions differed from those of the bioethicists. The most obvious differences concerned the standards and procedures for determining medical futility. The court had asked the ethicists to answer several questions regarding medical futility, including: How should the decision be made regarding what treatments are futile and under what standards, and who should make such decisions?

The bioethicists, describing the factual situation as "unambiguous," summarized:[47] "We conclude from the record that AMB falls clearly and non-controversially under the concept of medical futility."[48] Further, the bioethicists approved the reasoning process of the "emergency house counsel" at the hospital (Ms. Mahinske), noting that it "nicely corresponds with a proposal for hospital policy in making futility determinations from the Council on Ethical and Judicial Affairs of the American Medical Association."[49]

Judge Whitbeck, however, found the determination of futility to have been made on the basis of inadequate evidence; the actions and information relied on by the house counsel to have been deficient; and the family court's authorization for the hospital to withdraw life-sustaining treatment to have been erroneous. "Mahinske (the house counsel) had an obligation to investigate why she could rely on Dr. Delaney-Black's testimony as wholly authoritative. In other words, even if Mahinske did not or could not secure a second opinion from an independent physician, she should have developed the record so that it reflected why the family court could trust Dr. Delaney-Black's testimony completely."[50]

Referring to the family court referee, Judge Whitbeck concluded that the decision to authorize withdrawing life support was clear error. Again, the court pointed to the lack of a second medical opinion:

> Referee Schummer clearly considered Dr. Delaney-Black's opinion incontrovertible. In reality, Dr. Delaney-Black's opinion may have been uncontroverted simply because no other physician was called to testify. Although every other physician may have agreed completely with Dr. Delaney-Black, referee Schummer apparently did not even consider the possibility that baby Allison's diagnosis and prognosis might be debatable. Nor did he ask to hear testimony from anyone else who had seen baby Allison or was concerned about her.
>
> Certainly, the evidence on the record was clear. Dr. Delaney-Black's testimony directly supported referee Schummer's findings and recommendation and the ultimate "order." However, this evidence was not convincing. If baby Allison were still alive, we would remand this case to the family court for an evidentiary hearing so the family court could

develop a minimally acceptable record describing baby Allison's diagnosis and prognosis as viewed by others. If that were impossible or unnecessary, the family would have an opportunity to explain its conclusion. On the basis of this inadequate record, we simply cannot find convincing evidence to support a decision to authorize Children's Hospital to withdraw baby Allison's life support. Therefore, we conclude that this decision was clear error.[51]

Here we see an obvious disagreement between Judge Whitbeck and the bioethicists regarding the adequacy of the decision-making process in which the professionals involved with AMB had engaged. However, despite this disagreement, Judge Whitbeck described the bioethicists' brief as "thoughtful." The material on medical futility may have provided background information or an argument or analysis that the court could consider and to which it could implicitly respond. In addition, the brief's presentation of material that was in line with standard legal thinking, such as standards and procedures for determining parental competence, and for surrogate decision making, may have confirmed Judge Whitbeck's reasoning regarding their application to this unusual case. Even though the bioethicists' answers to several questions were ultimately rejected, the brief served as a dialog partner for the court.

4. When Are *Amicus Curiae* Briefs Unhelpful?

Although judges do not often cite bioethics briefs or note their "thoughtfulness," neither do they usually criticize them. An exception is Judge Richard Posner, who has criticized "The Philosopher's Brief," which was submitted by six prominent philosophers to the Supreme Court in the physician-assisted suicide cases. The brief was not cited in any of the Justices' opinions. Judge Posner's thesis regarding why the Philosopher's Brief was not cited—that moral theory is useless to law[52]—is highly controversial, and debate regarding that view is beyond the scope of this book. Nonetheless, two reasons Judge Posner offers in support of his belief that judges are reluctant to engage philosophical issues are relevant to the general question of the helpfulness of bioethics briefs: judges work under time pressures, making them reluctant to engage with esoteric arguments presented in *amicus curiae* briefs and judges want to preserve the autonomy of law rather than make it the handmaiden of moral philosophy.[53]

Judge Posner has complained elsewhere about the over-abundance of briefs submitted to courts. His view—a standard view—is that briefs that largely duplicate the positions and arguments advanced by the parties provide little or no assistance to judges. Explaining why he was denying a leave to file an *amicus curia* brief in 1997, Judge Posner wrote: "After 16 years of reading *amicus curiae* briefs, the vast majority of which have not assisted the judges, I have decided that it would be good to scrutinize these motions in a more careful, indeed a fish-eyed, fashion. The vast majority of *amicus curiae* briefs are filed by allies of the litigant and duplicate the arguments made in the litigants' briefs, in effect merely extending the length of the litigants' brief. Such *amicus* briefs should not be allowed."[54] Judge Posner is not alone in this view. Recall that Supreme Court Rule 37.1 warns that an *amicus curia* brief that does not bring material to the attention of the Court that is relevant and that the parties have not already brought, burdens the Court, and should not be filed.[55]

Judge Posner also claimed that judges are reluctant to engage philosophical issues because they want to preserve the autonomy of law rather than make it the handmaiden of moral philosophy. If this claim were true in an absolute sense, there would be no need for an examination, such as this one, regarding the uses of bioethics in decisional law; likewise, if judges had no interest in law's identity as law (rather than "law as bioethics"), there would also be no need for this examination. Because the claim is true under certain conditions, but not others, writers of bioethics *amicus curiae* must tread a fine line. On one hand, traditional legal arguments will already have been made by the parties. Judges may properly view repetition of those arguments in bioethics briefs as unnecessarily duplicative, burdensome, and, therefore, unwelcome. On the other hand, briefs that include only ethical arguments may instead be unhelpful because judges want to avoid unnecessarily blurring boundaries between law and ethics.

5. Summary

Bioethics testimony and briefs are potentially helpful to judicial reasoning. We have seen that a judge might view ethics testimony as helpful in finding legislative facts, as did Judge Hamilton when finding Constitutional facts in *Planned Parenthood Fed'n of Am. v. Ashcroft*. When facts are being adjudicated, providing help through bioethics testimony is more difficult, as we saw in *Hall v. Anwar* and *Schindler v. Schiavo*. A clear connection between the ethics testimony and the law's task, such as the link between testimony regarding ethical constraints and the constraints on production of evidence Bristol-Myers Squibb faced in *Zenith Lab. v. Bristol-Myers Squibb Co.*, is necessary. If bioethics testimony includes ethical standards that conflict with or distract from the law governing the case, however, the value of the testimony may be outweighed by the risk of confusion, as in *Biddison v. Facey* and *In re Rezulin Prods. Liab. Litig.* Finally, if bioethics experts challenge legal boundaries, as in *Rezulin* or *Bowman v. Corr. Corp. of Am.*, their contribution is unhelpful in litigation contexts.

Bioethics *amicus curiae* briefs can also be helpful. In *Washington v. Glucksberg*, the Brief for Bioethics Professors helped Justice Rehnquist make a claim regarding a slippery slope. The invited brief for *In re AMB* likely functioned as an implicit dialog partner for Judge Whitbeck. Bioethics briefs that are duplicative of arguments made by the parties or that do not observe law's boundaries, however, as Judge Posner apparently thought the Philosopher's Brief failed to do, are not helpful.

We have seen judges deal with the issue of moral pluralism in several cases. In *Planned Parenthood* v. *Ashcroft*, testimony regarding the range of moral views among professionals helped the court determine that an exception was necessary under *Stenberg* v. *Carhart*. In *Hall* v. *Anwar*, Judge Altenbernd voiced his conviction that the expert testimony was harmless to the legal process, but felt the need to address a potential First Amendment problem—that it was presented by a priest. Although law was open to the fact of moral diversity in *Planned Parenthood* v. *Ashcroft*, *Hall* v. *Anwar* signals, and *In re Rezulin Prods. Liab. Litig.* clearly illustrates, that law's receptivity is not easily extended to the norms themselves—especially to norms that seem to be religious norms or that are expressed in a pidgin of ethical–legal language.

Endnotes

[1]F. R. Evid. 702. Testimony by Experts: If scientific, technical, or other specialized knowledge will assist the trier of fact to understand the evidence or to determine a fact in issue, a witness qualified as an expert by knowledge, skill, experience, training, or education, may testify thereto in the form of an opinion or otherwise, if (1) the testimony is based upon sufficient facts or data, (2) the testimony is the product of reliable principles and methods, and (3) the witness has applied the principles and methods reliably to the facts of the case. As amended effective December 1, 2000.

[2]S. Ct. R. 37.1.

[3]Davis KC. An approach to problems of evidence in the administrative process. Harv L Rev 1942;55:364–425 at 402.

[4]The Federal Rules of Evidence govern the admissibility of evidence in the federal court system. A large majority of states have adopted similar or identical rules for us in their courts. This volume uses the federal rules.

[5]Advisory Committee Note to Federal Rule of Evidence 201(a), adopting Davis' terminology, describes legislative facts as "those which have relevance to legal reasoning and the lawmaking

process, in the formulation of a legal principle or ruling by a judge or court."

[6]Imwinkelried EJ. Expert testimony by ethicists: What should be the norm? J Law, Med & Ethics 2005;33:198–212; Spielman B. Bioethics testimony: Untangling the strands and testing their reliability. J Law Med Ethics 2005;33:222–233; Barnes DW. Imwinkelried's argument for normative ethical testimony. J Law, Med & Ethics 2005;33:234–241; Latham SR. Expert bioethics testimony. J Law, Med & Ethics 2005;33:242–247.

[7]Zenith Laboratories, Inc., Plaintiff, v. Bristol-Myers Squibb Co., Defendant. 1992 US Dist. LEXIS 11540 (1992).

[8]35 U.S.C. §§ 1 et seq.

[9]Agich G, Spielman B. Ethics expert testimony: Against the skeptics. J Med Philos 1997;22:381–403 at 396–397. Spielman B, Agich G. 1999. The future of bioethics testimony: guidelines for determining qualifications, reliability and helpfulness. San Diego L. Rev 1999;36:1043–1075.

[10]Planned Parenthood Federation of America, et al., Plaintiffs v. John Ashcroft, Attorney General of the United States, in his official capacity, Defendant. City and County of San Francisco, Plaintiff Intervenor, v. John Ashcroft, Attorney General of the United States, in his official capacity, Defendant, 320 F. Supp. 2d 957 (2004).

[11]Planned Parenthood v. Casey, 505 U.S. 833 (1992).

[12]Don Stenberg, Attorney General of Nebraska, et al. v. Leroy Carhart, 530 U.S. 914 (2000).

[13]Planned Parenthood Fed'n of Am. v. Ashcroft, 320 F. Supp. 2d 957, 1013–1014 (2004), citing Turner Broadcasting System, Inc., et al., Appellants v. Federal Communications Commission, et al., 520 U.S. 180, 196 (1997).

[14]Planned Parenthood Fed'n of Am. v. Ashcroft, 320 F. Supp. 2d 957, 1025 (2004).

[15]Planned Parenthood Fed'n of Am. v. Ashcroft, 320 F. Supp. 2d 957, 1009 (2004).

[16]Planned Parenthood Fed'n of Am. v. Ashcroft, 320 F. Supp. 2d 957, 1031 (2004).

[17]Fletcher JC. Bioethics in a local forum: Confessions of an "expert" witness. J Med Philos 1997;22:297–324.

[18]William Daubert, et ux., et al., Petitioners v. Merrell Dow Pharmaceuticals, Inc., 509 U.S. 579, 591 (1993).

[19]USCS F. R. Evid. 401 Rev. 2002.

[20]*In re* Guardianship of: Theresa Marie Schiavo, Incapacitated. Robert Schindler and Mary Schindler, Appellants, v. Michael Schiavo, as Guardian of the person of Theresa Marie Schiavo, Appellee, 780 So. 2d 176 (2001).

[21]Schindler v. Schiavo (*In re* Schiavo), 780 So. 2d 176, 179 (2001).

[22]James M. Hall and Betty Jo Hall, individually, and as the natural guardians of Larry Joseph Hall, a minor, and Joe Bohannon, as guardian of the property of Larry Joseph Hall, Appellants, v. M. Naveed Anwar, M.D., and DeSoto Memorial Hospital, Inc., d/b/a DeSoto Memorial Hospital, Appellees, 774 So. 2d 41 (2001).

[23]Hall v. Anwar, 774 So. 2d 41, 44 (2001).

[24]Hall v. Anwar, 774 So. 2d 41, 43 (2001).

[25]Agich G, Spielman B. Ethics expert testimony: Against the skeptics. J Med Philos 1997;22:381–403. Spielman B, Agich G. 1999. The future of bioethics testimony: guidelines for determining qualifications, reliability and helpfulness. San Diego L. Rev 1999;36: 1043–1075.

[26]Biddison v. Facey was settled. A related case is Facey Medical Group, et al., Petitioners v. Los Angeles County Superior Court, Respondent, Dorothy Biddison, et al., Real Parties in Interest, 1997 Cal. LEXIS 5080 (1997).

[27]Dorothy Biddison, James C. Biddison, etc., Plaintiffs, vs. Facey Medical Group, Khosro Sadeghani, M.D., Northridge Hospital, Niles Chapman, M.D., Take Care Health Plan, and Does 1-100, Inclusive, Defendants, Case No. PC016239X, deposition of John J. Paris, S.J., in Irvine, Cal. on June 17, 1998, p. 81 line 19 to p. 84 line 19.

[28]Dorothy Biddison, James C. Biddison, etc., Plaintiffs, vs. Facey Medical Group Khosro Sadeghani, M.D., Northridge Hospital, Niles Chapman, M.D., Take Care Health Plan, and Does 1-100, Inclusive, Defendants, Case No. PC016239X. Deposition of John J. Paris, S.J., in Irvine, Cal. (June 17, 1998) at p. 80–84. The transcript of the testimony refers to "Satewood's (ph.) opinion." The case to which the bioethicist is referring is Superintendent of Belchertown State School & another v. Joseph Saikewicz, 373 Mass. 728, 737 (1977), which was quoting a law review article, which in turn was quoting a medical journal: It has been said that "[t]he law always lags behind the most advanced thinking in every

area. It must wait until the theologians and the moral leaders and
events have created some common ground, some consensus."
Burger WE. The law and medical advances. Ann Intern Med
1967;67:Suppl 7:15–18 quoted in Elkinton JR. The dying patient,
the doctor, and the law. Vill L Rev 1968;13:740 at 743. "We therefore think it advisable to consider the framework of medical ethics
which influences a doctor's decision as to how to deal with the terminally ill patients. While these considerations are not controlling,
they ought to be considered for the insights they give us." The court
in Superintendent of Belchertown State School v. Saikewicz
clearly understood itself to be making law, and understood the relationship of ethics to its task differently than the bioethics expert in
Facey Medical Group v. Los Angeles County Superior Court did.
The court wrote: "We recognize at the outset that this case presents
novel issues of fundamental importance that should not be resolved
by mechanical reliance on legal doctrine. The task of establishing
a framework in the law on which the activities of health care personnel and other persons can find support is furthered by seeking
the collective guidance of those in health care, moral ethics, philosophy, and other disciplines." Superintendent of Belchertown State
School v. Saikewicz, 373 Mass. 728, 736 (1977).

[29]F. R. Evid 403.

[30]Andre J. Bioethics as Practice. Chapel Hill and London: The
University of North Carolina Press, 2002.

[31]*In re*: Rezulin Product Liability Litigation (MDL No. 1348), 309 F.
Supp. 2d 531 (2004).

[32]*In re* Rezulin Prods. Liab. Litig., 309 F. Supp. 2d 531, 545 (2004).

[33]*In re* Rezulin Prods. Liab. Litig., 309 F. Supp. 2d 531, 558 (2004).

[34]*In re* Rezulin Prods. Liab. Litig., 309 F. Supp. 2d 531, 558 footnote 102
(2004). The potential confusion in sorting ethics language from
legal language was addressed head-on in a pretrial decision by New
Jersey Superior Court Judge Higbee in 2005, mentioned in the
introduction to this book. The purpose of the trial was to determine
whether Merck, which manufactured the COX-2 inhibitor Vioxx,
was liable for the death of a heart attack victim. Judge Higbee
barred the use of the words "ethics" and "morality" in discussions
of medical evidence in order to limit "subjective" and "inflamma-
tory" comments. Higbee urged, "Let's keep it on a scientific and

factual level ... I just don't want the issue of what's ethical—it's not a legal phrase." She added, "That's not to say the trial isn't about what's wrong or right. That's what the law is about." (No author listed). Vioxx: New Jersey Judge Bars Testimony on 'Ethics' 'Morality.' American Health Line. September 9, 2005 (No volume, issue, or page numbers listed). Available at LEXIS NEXIS.

[35]United States of America, Apellee, v. Paul A. Bilzerian, Defendant–Apellant, 926 F.2d 1285, 1294 (1991).

[36]*See* 4 Weinstein's Federal Evidence § 702.03 [2][a].

[37]Patricia Bowman, Plaintiff–Appellee/Cross–Appellant, v. Corrections Corporation of America, Defendant–Appellant/Cross–Appellee, 350 F.3d 537, 547 (2003).

[38]Bowman v. Corr. Corp. of Am., 350 F.3d 537, 877 (2003).

[39]Bruce Ennis describes three kinds of help that *amicus* briefs can provide to parties: they can help the party flesh out arguments the party is forced to make in summary form because of page limits and other considerations; they can make arguments the party wants to make but cannot make itself for political or tactical reasons or because they lack credibility on the issue; and they can inform the court of the broader public interest involved, or of the broader implications of a ruling. Ennis BJ. Symposium on Supreme Court Advocacy: Effective *Amicus* Briefs. The Catholic University Law Review 1984;33:603–609 at 605–608.

[40]1995 US Briefs 1858; 1996 US S. Ct. Briefs LEXIS 740.

[41]*In re* AMB, Minor: Brief of Scholars in Medical Ethics as Amici Curiae, Court of Appeals Docket No. 218869, 1–36, June 26, 2001.

[42]Washington v. Glucksberg at 521 U.S. 733; 117 S. Ct. 2274.

[43]Compassion in Dying v. Washington, 79 F.3d 790 at 831.

[44]Brief for Bioethics Professors at 32.

[45]*See* Rich B. Strange bedfellows: How medical jurisprudence has influenced medical ethics and medical practice. New York: Kluwer Academic, 2002 at 161 for a critique of this assertion.

[46]In the Matter of AMB, Minor, Family Independence Agency, Petitioner–Appellee, v AMB, Family Division, Respondent–Appellant, 248 Mich. App. 144 (2001).

[47]*In re* AMB, Minor: Brief of Scholars in Medical Ethics as *Amici Curiae*, Court of Appeals Docket No. 218869, 1–36, at 33. June 26, 2001.

[48]*In re* AMB, Minor: Brief of Scholars in Medical Ethics as *Amici Curiae*, Court of Appeals Docket No. 218869, 1–36, at p. 30. June 26, 2001.

[49]*In re* AMB, Minor: Brief of Scholars in Medical Ethics as *Amici Curiae*, Court of Appeals Docket No. 218869, 1–36, at 31. June 26, 2001.

[50]Family Independence Agency v. AMB (*In re* AMB), 248 Mich. App. 144 at 233 (2001).

[51]Family Independence Agency v. AMB (*In re* AMB), 248 Mich. App. 144 at 215–216 (2001).

[52]Richard A. Posner, The Problematics of Moral and Legal Theory (Cambridge, MA: The Belknap Press of Harvard University Press, 1999) at 130–134. Ben Rich has ably analyzed the bioethics briefs submitted in the assisted suicide cases. Ben Rich, Strange Bedfellows, at 157–162.

[53]Richard A. Posner, The Problematics of Moral and Legal Theory (Cambridge, MA: The Belknap Press of Harvard University Press, 1999) at 132.

[54]John H. Ryan, Petitioner, v. Commodity Futures Trading Commission, Respondent, 125 F.3d 1062 (1997). (Posner, J.).

[55]S. Ct. R. 37.1.

2   Health Care Ethics Committee Determinations

In 1975, a *Baylor Law Review* article recommended using hospital ethics committees for end-of-life decision making.[1] Its author, Dr. Karen Teel, thought that "such an entity could lend itself well to an assumption of a legal status which would allow courses of action not now undertaken because of the concern for liability." A year later, the New Jersey Supreme Court, considering whether to permit withdrawal of ventilator support from Karen Ann Quinlan, adopted Teel's idea and endorsed ethics committees.[2] The New Jersey court's endorsement provided a crucial boost to the fledgling ethics committee movement, which has since become widespread in the United States. Most health care delivery organizations now have health care ethics committees (HECs), which make determinations, or recommendations, on request, regarding medical treatment in difficult clinical cases.

Unlike expert testimony and bioethics *amicus* briefs, HEC determinations are developed for use in clinical settings. Nevertheless, using HECs to influence the legal system has been a goal of not only Dr. Teel, but also of many HEC proponents. This chapter explores the interactions between HECs and law,

From: *Bioethics in Law*
By: B. J. Spielman © Humana Press Inc., Totowa, NJ

specifically the role of HEC determinations in judicial reasoning. Do judges treat HEC work as adjudicative fact? As legislative fact? Do they treat HEC determinations as normative facts?

1. HEC Determinations and Adjudicative Fact

Judges often treat HEC determinations as adjudicative facts. Some clinical or research cases have already been reviewed by an HEC, and a court treats this review as one of the facts of the case. Another potential use of HECs—as a source of other adjudicative facts regarding the case—is suggested in *Abdullah* v. *Pfizer*.[3] After an outbreak of cholera, meningitis, and gastroenteritis in Nigeria, a putative class action suit sought redress from Pfizer Pharmaceuticals for injuries arising from the experimental administration of an antibiotic. The case was initially filed in Connecticut, where Pfizer's Global Research and Development World Headquarters was located, and was subsequently transferred to federal court in New York. Plaintiffs brought an action under the Alien Tort Statute because Pfizer purportedly violated not only Food and Drug Administration regulations, but also the Nuremburg Code, the Declaration of Helsinki, the International Covenant on Civil and Political Rights, and customary international law.[4] Plaintiffs in a parallel case had alleged corruption and bias in the Nigerian judiciary. Pfizer made a motion to dismiss on several grounds, including forum *non conveniens*, a discretionary device permitting a court to dismiss a claim if the inconvenience to the defendant of the forum chosen by the plaintiff is out of proportion to its convenience for the plaintiff.

District Judge Pauley III decided that, even if his court had had subject matter jurisdiction, he would dismiss on the ground of forum *non conveniens*, noting that even the plaintiffs would find a Nigerian forum helpful. They would have to rely on local Nigerian hospitals, governmental officials, and injured persons to establish causation, injury, and damages. Interestingly, he noted that plaintiffs would also rely on the treating hospital's ethics committee in Nigeria to obtain "knowledge of the relevant events." That is, he expected that the HEC could provide some of the "who, what, when, and where, and how" of the disputed events. The HEC could not do so, however, because it did not exist when the research began. Pfizer, whose actions created international controversy, had submitted to the Food and Drug Administration a letter of approval from the HEC that had been backdated to precede the start of the research.

2. HEC Determinations as Legislative Facts

Occasionally a judge will use an HEC's work—whether consistent or inconsistent with core legal norms—to represent the way HECs function. The actions of a Wisconsin HEC became such legislative facts, used by a Wisconsin Supreme Court Justice to inform (or warn) future judges regarding how HECs can actually work.

In 1992, the Wisconsin Supreme Court had considered the case of *In re Guardianship of L.W.*, a 79-year-old man who lay in a persistent vegetative state.[5] The trial court had set forth 12 criteria in its memorandum opinion to guide guardians who were determining whether forgoing life-sustaining treatment would be in the best interests of their wards. One of the criteria was "the recommendation, if any of a bioethics committee."[6] The Supreme Court said that it was not adopting all 12 criteria but was suggesting it was "an option" to consider the advice of bioethics committees.[7] The court went on to suggest that, "if [a bioethics committee] is available, the guardian should request it to review the decision, and should consider its opinion in determining whether it is in the patient's best interests to forego treatment."[8] Further, the court thought that a right to refuse treatment for incompetent individuals, if consistent with medical ethics as represented by an HEC, might serve a norm-enforcing function by protecting the integrity of the medical profession:

The state's interest in protecting the integrity of the medical profession is not implicated in this case. *In re Guardianship of L.W.*'s physicians initiated the action by conditionally (i.e., if L.W.'s condition remained unchanged for another 4 weeks) requesting the guardian's consent to withdraw treatment. Their actions were consistent with current medical ethics in so far as approval was sought and given by the Bioethics Committee of Franciscan Health System. Current Opinions of the Council on Ethical and Judicial Affairs of the American Medical Association 2.18, "Withholding or Withdrawing Life-Prolonging Medical Treatment" (1986); Position of the American Academy of Neurology on Certain Aspects of the Care and Management of the Persistent Vegetative State Patient, 39 Neurology 125 (1989). Thus, a decision to withhold or withdraw treatment will not impugn the integrity of the profession. Indeed, the existence of a protected right to refuse treatment for all individuals, competent or incompetent, may, in a sense, protect the integrity of the medical profession. In the absence of such a protected right, physicians may be discouraged from attempting certain life-sustaining medical procedures in the first place, knowing that once connected they may never be removed. Conroy, 98 N.J. at 370, 486 A.2d at 1234. The existence of this right will prevent premature and rash decisions to allow a patient to die, and will remove the potential conflict for the medical profession between ordinary compassion and the Hippocratic Oath.[9]

The Wisconsin court had optimistically carved out a potentially significant role for HECs. HECs were understood to be guardians of the integrity of the medical profession. However, 5 years later, when the Court was considering the case of *Spahn* v. *Eisenberg* (*In re Edna M.F.*), the efforts of a different HEC, which were not consistent with legal norms, were viewed much less favorably, but still treated as legislative fact. Concurring with the court's decision not to permit forgoing life-sustaining treatment from the 71-year-old Alzheimer's disease patient who was bedridden but not unconscious, Chief Justice Abramson wrote separately to clarify "the majority opinion's characterization of several aspects of the controlling case in the majority's decision, L.W...." including the role the court had carved out for HECs:

L.W. commented favorably on the role of the health care provider's ethics committee. Hospital or nursing home ethics committees provide an important forum for careful deliberation regarding the decision to withhold life-sustaining medical treatment. Based on the limited record before us, it seems that the committee reviewing the request by Ms. F.'s guardian did not function effectively. Had Ms. F. been in a persistent vegetative state and had an interested person objected to the withdrawal of nutrition, the circuit court stated that it would have been unable to give weight to the committee's purported determination that withholding of nutrition was the ethically proper course. The circuit court noted that no formal minutes or report of the meeting was produced at the hearing and that the committee members apparently functioned without either a shared body of rules or training in ethics. In fairness to the committee members in this case, it must be noted that the committee had only recently been formed and had deliberated in perhaps only one other case.

The circuit court also seemed troubled, as am I, with the apparent focus of the ethics committee's investigation. The committee seemed to understand that its function was to reach a determination that would insulate the facility from legal liability rather than the determination that best comported with medical ethics. The focus of all participants in this fateful and difficult process should be on the propriety of taking action that will lead to a person's death. The health care facility's liability concerns must not be allowed to interfere with the guardian's efforts to assure the exercise of the ward's right to be free of unwanted life-sustaining medical treatment if the guardian has determined, in consultation with the physicians, that the ward is in a persistent vegetative state and it is in the ward's best interests to with-hold such treatment.[10]

The Justice wrote that her comments regarding the HEC were necessary, because "further discussion of the application of *L.W.* to the present case is needed."[11] Although *L.W.* did not explicitly require an evaluation, she wanted courts to evaluate the work of HECs before giving them any normative weight. What she expected was that the HEC would practice procedural fairness, including functioning under an explicit set of rules, and that it would protect individual rights, including a ward's right to be free of unwanted life-sustaining medical treatment. If the HEC failed to meet these expectations, Justice Abramson wanted to limit future courts' normative uses of HEC determinations.

Unlike the Wisconsin Supreme Court, the Kentucky Supreme Court had, by the early 1990s, already anticipated the issue of HEC work that did not conform to core legal norms. In its *DeGrella* decision, the Kentucky court had given ethics committees veto power in nursing home residents' treatment decisions.[12] At the same time, the court recognized that not every such decision would be consistent with core legal norms. It had stated:

> If the attending physician, the hospital or nursing home ethics committee where the patient resides, and the legal guardian or next of kin, all agree and document the patient's wishes and the patient's condition, and if no one disputes their decision, no court order is required to proceed to carry out the patient's wishes.... [However, a] false or fraudulent, and collusive, decision is beyond the power of a court to approve before or after the termination of life-sustaining medical treatment.[13]

The Kentucky court had made clear in DeGrella that there would be an exception to its otherwise deferential posture toward HEC determinations if HECs violated basic legal norms. Because this exception had been established, the *Woods* v. *Commonwealth* court had no reason to limit the scope of the HEC role, even if

HEC work was poor (which the court did not think it was, in this case).[14] In fact, the court expanded the role of HECs in *Woods*, viewing them as norm-enforcing institutions that could substitute for judicial oversight. In *Woods*, the existence of HECs, properly constrained by core legal norms, was, thus, used as a legislative fact to persuade the guardian and the public that the statute would be safely implemented.

3. HEC Determinations as Normative Fact

For several reasons, HEC determinations can have greater potential than other bioethics materials to influence judicial reasoning. First, HEC determinations are, similar to trial court determinations, specific to a particular case; they address roughly the same set of facts and provide normative guidance for them. Second, the HEC's determination is rarely opposed by another HEC's determination.[15] These factors create the potential for judges to be more receptive to, and perhaps less critical of, HEC determinations than of other bioethics communications. Third, in several jurisdictions, as Dr. Teel hoped, legislatures or judges have assigned to HECs a legal role. Thus far, Davis' distinction between adjudicative fact and legislative fact has been used to analyze interactions between HEC communications and law. If a judge admits an HEC determination, however, the judge may also give it normative weight—in effect, treating it as normative fact.

In an especially contentious case from California, for example, the actions of Florence Wendland, the mother of Robert Wendland, suggest a suspicion that an HEC determination might be treated normatively by a judge. The facts and procedural history of the Robert Wendland case are complicated. However, to understand Florence Wendland's objection to admitting an HEC determination without examining or cross-examining the HEC's members, the following summary will suffice.[16]

Robert Wendland was a middle-aged man who was left severely brain damaged by a motor vehicle accident in 1993.

He was conscious and sometimes able to respond to simple commands, but was unable to speak, and was completely dependent on others for his care. At the time the California appellate court decided his case in 2000, he was receiving nutrition and hydration through a feeding tube. A 20-member ethics committee determined that it had no objection when his wife, who was also his conservator, ordered withdrawal of nutrition and hydration.

After learning of the decision to withdraw treatment, Florence challenged her daughter-in-law's conservatorship and subpoenaed all of the members of the ethics committee. Her subpoenas were quashed by the probate court. Florence then complained that, because her subpoenas had been quashed, the court of appeals of California "should not give any weight whatsoever to the committee's decision."[17]

The HEC's determination could have been used for a number of purposes, one of which would be to guide the court's normative decision making. Florence wanted to make sure that the unchallenged HEC determination was not treated as normative fact. The court declined, however, to reject the evidence of the HEC's determination. It wrote, "Even assuming for the sake of argument the trial court erroneously quashed the subpoenas (a matter we do not decide), we see no basis for rejecting the evidence on this issue adduced at trial."[17] In other words, the HEC determination could become at least adjudicative fact. But, after a discussion of the HEC's work, the appellate court noted that Florence had learned through an anonymous phone call of the plan to remove the tube.[17] By calling attention to the HEC's omission, the court hinted that it might not follow the HEC's guidance.

Florence Wendland's fear that the HEC determination might be treated normatively did not materialize. On further appeal, the Supreme Court of California, too, noted that Florence had learned through an anonymous phone call about the plan to remove Robert's feeding tube, and added that the HEC had not spoken with Robert's mother.[18] Implicit in the judge's inclusion of these procedural omissions is disapproval of the HEC's work. The high court ultimately decided, against the recommendation of the HEC, that nutrition and hydration should not be withdrawn. Was Florence Wendland's concern completely unfounded? Do judges ever treat HEC recommendations normatively? Do they ever view the recommendations uncritically?

3.1. When Legislatively Required

Occasionally, judges are forced to treat HEC determinations as normative fact. Legislatures or regulatory agencies can assign a normative task to HECs by statute or regulation. When they do, the normative role that HECs have in the clinical setting carries over into judicial reasoning. This potential was made clear in a concurring opinion in the Texas case, *Nikolouzos* v. *St. Luke's Episcopal Hosp.*[19]

Texas has a "futility law"—a set of procedures enabling health care providers, without fear of liability, to refuse to provide life-sustaining treatment that patients and/or their families request. Under that law, when a patient has directed an attending physician to give life-sustaining treatment that the physician thinks is inappropriate, the physician may ask the HEC to make a determination regarding the appropriateness of the treatment. This is a step in the statutorily outlined process of either transferring the patient to another facility that will provide the treatment or letting the patient die without the treatment in the facility using the procedure.[20]

Consistent with the statute, a Texas HEC had determined that life-sustaining treatment requested by the family for Mr. Nikolouzos was "inappropriate."[21] A state appellate court dismissed, for lack of jurisdiction, an interlocutory appeal from a denial of two applications by the family for temporary restraining orders against the hospital. In a concurring opinion, however, Judge Fowler explained that, if the court had had jurisdiction, the HEC determination would have been a reason to exclude from consideration a physician's report stating that Mr. Nikolouzos was not brain dead.[22] The Nikolouzos' had offered that report in the hope of extending Mr. Nikolouzos' life support. Judge Fowler wrote: "As for the proof already before the court from Dr. John Meyer that Mr. Nikolouzos was not brain dead, St. Luke's has pointed out, and the trial judge found, this evidence was irrelevant to the issue before the court. *Section 166.046* permits the withdrawal of life-sustaining care for patients who are not brain dead if the hospital's ethics committee has determined the care is inappropriate."[23]

Because HEC determinations had been assigned a normative task by Texas legislature, the HEC determination became for Judge Fowler an exclusionary reason—a reason that excluded certain facts regarding the status of Mr. Fowler's brain from the decision-making process.[24] Even if the court had found jurisdiction, the fact that Mr. Nikolouzos' brain was alive could not have been considered by the court; the HEC determination precluded it.

In Judge Fowler's concurrence, the HEC determination of the "inappropriateness" of medical treatment is, thus, more than an adjudicative fact. It also has normative weight that she hypothesized would guide judicial reasoning regarding an evidentiary matter, if the court had jurisdiction. Judge Fowler seemed to want to assure the family that the HEC determination was not the only factor preventing continuation of life support. The Nikolouzos family had not proven "by a preponderance of the evidence, that there is a reasonable expectation that a physician or health care facility that will honor the patient's directive will be found if the time extension is granted."[25] Because this proof that an alternative facility would cooperate was required by statute, Judge Fowler expected that the Nikolouzoses could not have prevailed on the merits.[26] The HEC determination, thus, was a normative fact in Judge Fowler's reasoning regarding a hypothetical in which the court would have had jurisdiction. The HEC determination would have been a reason not to honor the Fowler's directive. Because of a legislative decision, the HEC determination's normative character would carry over into judicial reasoning.

3.2. At Judicial Discretion, When Not Inconsistent With Legal Norms

If judges are not required to treat HEC determinations normatively, they assess whether HEC work supports core legal norms in deciding whether to give HEC work normative weight. Sometimes they approve of the work the HEC has performed, and sometimes they find that it has not used fair processes or that it violates individual rights. A positive assessment often corresponds to assigning normative weight to the HEC determination.

In the Matter of AB is a case in which a judge assessed the HEC's work, approved of it, and used it in her reasoning. A 3½-year-old child lay in a persistent vegetative state in a New York hospital.[27] The child's mother wanted to withdraw ventilator support. The child's father, who was separated from the mother, supported the decision. Hospital policy did not permit forgoing treatment in this kind of case. The mother asked the court to rule that she had the authority to remove AB from the ventilator.

New York had no law directly on point. Judge Ling-Cohan relied, by analogy, on New York's Health Care Decisions Act, state public health law, and state case law to support the proposition that AB's mother had the authority to make a best interests decision to forgo life-sustaining treatment for her child. The mother had discussed the decision with family members, including the child's father, who agreed with her.

The judge acknowledged what the HEC had done, both substantively and procedurally. In particular, she noted that the HEC had guided AB's mother through an analysis similar to the best interests analysis articulated in New York state law. The HEC had met numerous times with the mother and others.[28] Partly on the basis of those meetings, Judge Ling-Cohan came to the conclusion that AB's mother's decision making was well-informed, and found clear and convincing evidence that it was in the child's best interests to forgo treatment.[29]

Other judges have noted both procedural and substantive problems in an HEC's work. In *Rideout* v. *Hershey Medical Ctr.*, the parents of a 2-year-old girl vehemently protested forgoing life-sustaining treatment that the hospital "through its ethics committee [had determined] was an appropriate step."[30] The parents claimed they had been assured that the ventilator would not be turned off in their absence. Allegedly, however, while they were in another part of the hospital arranging to obtain legal help, they were informed over the intercom that their daughter's ventilator was being withdrawn. The Pennsylvania court of common pleas would decide whether to dismiss the hospital's objections to the parents' complaints that the hospital had violated their rights and those of the child under state and federal law, including the First and Fourteenth Amendments.[31] In overruling the hospital's objections, Judge Turgeon scolded the hospital for failing to protect individual rights. The court wrote:

> This case is truly exceptional, in that the hospital here unilaterally asserted, and in fact usurped, the minor incompetent's state privacy and/or federal liberty-based right to refuse life-sustaining medical treatment. In contrast, the Rideouts attempted to act, albeit too late, in the role traditionally asserted by the state, which is to act to preserve human life.[32]

Far from accepting the HEC determination uncritically, Judge Turgeon implied that the HEC's work, as part of the hospital's efforts to end life-sustaining treatment, contravened core legal norms. Responding to the HEC's and hospital's disregard for the family's moral views, Judge Turgeon declined to dismiss the parents' First Amendment and other claims.

4. Summary

Twenty years ago, when only a few HEC determinations had made their way into US courts, legal scholar Susan Wolf asked:

> Have courts chosen to defer to committee determinations or to ignore them?...Do the committees' determinations tip the scales of justice? Do the courts regard committees as better suited than the courts to decide, and so overturn committee determinations only in rare circumstances?

Wolf recommended that courts admit HEC determinations and then evaluate them on a case-by-case basis to decide what weight each deserves.[33]

Courts seem to be moving in the direction that Wolf recommended. Judges not only treat HEC review as adjudicative fact and, in rare instances, expect HECs to provide other adjudicative facts, such as in *Abdullah* v. *Pfizer*, but they also admit HEC determinations and then evaluate them on a case-by-case basis to decide whether each determination deserves normative weight. Courts have not automatically given HEC work normative weight unless a legislature, regulatory agency, or higher court requires them to do so, as illustrated in Judge Fowler's hypothetical in *Nikolouzos* v. *St. Luke's Episcopal Hosp.* Judges do not hesitate to criticize or reject HEC determinations that violate either law's core procedural norms, as *Spahn* v. *Eisenberg* illustrates, or law's core substantive norms, as *Rideout* v. *Hershey Medical Ct.* illustrates. However, HEC determinations may be given a normative role in judicial reasoning, as *In re AB* illustrates, if those determinations support core legal norms.

Judges use their assessments of HECs to make or modify law, as in *Spahn* v. *Eisenberg* (*In re Edna M.F.*), in which the unacceptable performance of an HEC was used as a reason to qualify the court's previous decision in *Guardianship of L.W.*, and in *Woods* v. *Commonwealth*, in which the expected role of

HECs as potential norm enforcers was a legislative fact that explained why upholding the challenged law would not result in abuse.

In Chapter 3, the determinations of institutional review boards are examined. Similar to HEC determinations, institutional review board determinations are made for purposes other than litigation and are used in a variety of ways in law.

Endnotes

[1]Teel K. The physician's dilemma: A doctor's view: What the law should be. Baylor L Rev 1975;27:6–9.

[2]In The Matter of Karen Quinlan, an Alleged Incompetent, 70 N.J. 10, 48–49 (1976).

[3]Abdullah v. Pfizer, 399 F. Supp. 2d 495 at 505.

[4]28 U.S.C. §1350.

[5]In the Matter of Guardianship of L.W., Incompetent: Paul J. Lenz, as Guardian Ad Litem, Appellant–Cross Respondent, v. L.E. Phillips Career Development Center, Guardian, Respondent–Cross Appellant, Eau Claire County, Respondent, and St. Francis Hospital, Respondent–Cross Appellant, 167 Wis. 2d 53, 98 (1992).

[6]*In re* Guardianship of L.W., 167 Wis. 2d 53, 66 (1992).

[7]*In re* Guardianship of L.W., 167 Wis. 2d 53, 73–74 (1992).

[8]*In re* Guardianship of L.W., 167 Wis. 2d 53, 89 (1992).

[9]*In re* Guardianship of L.W., 167 Wis. 2d 53 (1992).

[10]Spahn v. Eisenberg (*In re* Edna M.F.), 210 Wis. 2d 557 (1997).

[11]Spahn v. Eisenberg (*In re* Edna M.F.), 210 Wis. 2d 557, 573 (1997).

[12]Martha Sue DeGrella, by and through her Guardian Ad Litem, Homer Parrent, III, Appellant v. Joseph G. Elston, Appelle, 858 S.W.2d 698, 710 (1993).

[13]DeGrella v. Elston, 858 S.W.2d 698, 710 (1993).

[14]The Kentucky Supreme Court mentioned that the HEC had unanimously recommended that treatment be foregone. However, the question before the court was the constitutionality of the Kentucky statute, not Mr. Woods' treatment (he had died during the course of litigation).

[15]One exception is the Torres case, in which several ethics committees were invited to give their opinion. *In re* Conservatorship of Rudolfo Torres. NW2d 1984;357:332–341. Sup Ct of Minn.

[16]*See also* FlorenceWendland, et al., Petitioners, v. The Superior Court of San Joaquin County, Respondent; Rose Wendland, Real Party In Interest, 49 Cal. App. 4th 44(1996); Conservatorship of the Person of Robert Wendland Rose Wendland, Petitioner and Appellant, v. Florence Wendland, et al., Objectors and Respondents; Robert Wendland, Appellant, 78 Cal. App. 4th 517 (2000); Conservatorship of the Person of Robert Wendland, Rose Wendland, etc., Appellant v. Florence Wendland, et al., Respondents. Robert Wendland, Appellant; 2 P.3d 1065 (2000); Conservatorship of the Person of Robert Wendland, Rose Wendland, as conservator, etc., Petitioner and Appellant, v. Florence Wendland, et al., Objectors and Respondents; Robert Wendland, Appellant, 26 Cal. 4th 519 (2001); Conservatorship of the Person of Robert Wendland; Rose Wendland, Appellant v. Florence Wendland, Respondent, 2001 Cal. LEXIS 6484 (2001).

[17]Conservatorship of the Person of Robert Wendland Rose Wendland, Petitioner and Appellant, v. Florence Wendland, et al., Objectors and Respondents; Robert Wendland, Appellant, 78 Cal. App. 4th 517, 527 note 8 (2000).

[18]Conservatorship of the Person of Robert Wendland, Rose Wendland, as Conservator, etc., Petitioner and Appellant, v. Florence Wendland, et al., Objectors and Respondents; Robert Wendland, Appellant, 26 Cal 4th 519, 525 (2000).

[19]Spiro Nikolouzos by his wife, Jannette Nikolouzos, Appellant v. St. Luke's Episcopal Hospital, Appellee, 162 S.W.3d 678 (2005).

[20]Texas Health & Safety Code Ann. § 166.046 (Texas Statutes and Codes annotated by LEXIS NEXIS (R) (2005).

[21]Nikolouzos v. St. Luke's Episcopal Hosp., 162 S.W.3d 678 (2005).

[22]Nikolouzos v. St. Luke's Episcopal Hosp., 162 S.W.3d 678 (2005). Justice Wanda K. Fowler concurring opinion at 682.

[23]Nikolouzos v. St. Luke's Episcopal Hosp., 162 S.W.3d 678, 683 (2005).

[24]Raz J. Practical reason and norms. London: Hutchinson & Co., 1975 at 47–48.

[25]Nikolouzos v. St. Luke's Episcopal Hosp., 162 S.W.3d 678, 683 (2005), Justice Wanda K. Fowler concurring opinion at 683.

[26]Nikolouzos v. St. Luke's Episcopal Hosp., 162 S.W.3d 678, 683 (2005) Justice Wanda K. Fowler concurring opinion at 682.

[27]In the Matter of AB, an Infant, by Her Mother, CD, Petitioner, 196 Misc. 2d 940, (2003).

[28]*In re* AB, 196 Misc. 2d 940, 960 (2003).

[29]*In re* AB, 196 Misc. 2d 940, 960–961 (2003).

[30]Rideout v. Hershey Medical Ctr., 30 Pa. D. & C.4th 57, 60 (1995).

[31]When the hospital unilaterally decide to remove life-sustaining treatment from a child, the plaintiff parents and plaintiff estate filed complaints against the hospital for negligent and intentional infliction of emotional distress and for the violation of their common law rights, constitutional rights, and rights under 42 U.S.C.S. Sec. 1395 and 29 U.S.C.S. Sec. 794. The hospital filed preliminary objections to the complaint.

[32]Rideout v. Hershey Medical Ctr., 30 Pa. D. & C.4th 57 (1995).

[33]Wolf S. Ethics committees in the courts. Hastings Cent Rep 1986;16:12–15 at 12.

3 Institutional Review Board Determinations

Institutional review boards (IRBs) were created pursuant to the 1974 National Research Act[1] to ensure that the rights and welfare of human subjects would be protected. The Act and implementing regulations were a response to research scandals, such as the Tuskegee Syphilis study, in which black men with syphilis were left untreated for more than three decades by the US Public Health Service, and the Jewish Chronic Disease Hospital study, in which live cancer cells were injected into hospital patients without consent. The regulations delegated to IRBs the responsibility to review and oversee research on human subjects. Although they would be highly regulated, law would rely on them in an ongoing manner. Susan Wolf describes IRBs as a "perfect example of a body conceived to do both law and ethics. They are required to apply the federal regulations, which are law, but those regulations are so open-textured and the overriding mission of IRBs is so clearly to protect human subjects, that IRBs must do ethics too."[2]

IRB recommendations are developed for a particular study and a particular set of researchers—not for litigation purposes. Those recommendations may include whether a study involving

From: *Bioethics in Law*
By: B. J. Spielman © Humana Press Inc., Totowa, NJ

human subjects could ethically proceed, whether potential research subjects are "at risk," and if so, whether the risks outweigh the potential benefits to the research subjects and the importance of the knowledge that might be gained from the research; what information should be disclosed in the informed consent process; whether the selection of subjects is equitable; what methods should be used for protecting confidentiality; whether disclosures to subjects regarding confidentiality are adequate; how the data will be monitored to ensure the safety of subjects; and whether incentives to participate can be offered, and, if so, the conditions under which the offer may be made.[3] Today, any of these bioethics recommendations, as well as communications regarding the processes by which they were reached, may interact with law.

During their first decade, IRBs were not challenged to publicly defend their decisions. However, an IRB determination was eventually challenged in court; concerns regarding how open judges would be to IRB judgments, and whether IRB determinations would be given normative weight surfaced. In *Head* v. *Colloton*,[4] a leukemia patient sought an injunction requiring the University of Iowa Hospital to send a letter of request to a potential bone marrow donor. The IRB had already mailed one request, and the recipient had declined to participate in the bone marrow registry study. Mr. Head wanted a second letter sent, but the IRB declined to do so. Mr. Head attempted to convince a judge that the University of Iowa Hospital IRB's decision making had been "arbitrary and capricious," and not based on any clear procedural guidelines or ethical norms.

Commenting on the case, Lidz, Meisel, and Roth argued that courts should not overturn the Iowa IRB's work, because doing so would undermine IRB authority and discourage IRBs from making difficult choices.[5] Implicit in this position is that courts should defer to IRBs' normative work. *Head* v. *Colloton* was ultimately decided by the Iowa Supreme Court, on the basis of the state's public records act rather than on the basis of the adequacy of the IRB's work. The judges neither overruled the IRB's determination nor referred to the IRB's determination as anything other than one of the facts of the case when it reversed the trial court's decision to open the records. Judge McCormick, writing for the court, simply noted that "When the hospital established the new program, its institutional review board approved a procedure for contacting persons listed on the registry to determine whether they would act as donors."[6]

However, questions regarding the role of IRB work in judges' practical reasoning have not disappeared; in fact, they have intensified along with litigation involving IRBs. Because they were conceived to do both law and ethics, their normative use is central, and that is where we begin. In this chapter, we ask how judges reason about the work of IRBs in contexts, such as suits against researchers, suits by researchers against IRBs regarding suspension of research projects, and disputes regarding health insurance coverage for IRB-approved protocols. Do judges overturn IRB work, as the commentators on *Head* v. *Colloton* feared they would? Do they treat IRB determinations as adjudicative fact? Do they treat them as normative fact or unreflectively translate them into law? Do they treat IRBs as if they could do both law and ethics?

1. IRB Determinations as Normative Fact

IRB determinations are often treated straightforwardly as facts of the case, as they were in *Head* v. *Colloton*. By virtue of their role in federal research law,[7] however, IRB determinations are often treated as normative fact in judicial reasoning.

Federal law authorizes IRBs to protect the public from harm. In some cases, this authority is exercised by preventing research from going forward, or by suspending research once it has begun. Such determinations can give rise to litigation by researchers against IRBs. In this type of case, judges sometimes assign normative weight to the IRB determination. For example, in *Halikas* v. *Univ. of Minn.*,[8] the IRB's supervisory actions were at issue. Dr. Halikas was a tenured professor and psychiatrist at the University of Minnesota Medical School. The IRB had approved his study of gamma-hydroxybutyrate as an aid in combating opium, cocaine, and methadone/heroin addition. The IRB received a letter from another physician/faculty member questioning whether consent of the subjects, most of whom were from Southeast Asia, was voluntary and informed, as well as whether the subjects were offered standard methadone treatment. Shortly thereafter, Dr. Halikas notified the IRB that he had canceled the study.

The IRB then began investigating the study, and notified the Food and Drug Administration (FDA) of its investigation. The university issued a press release indicating that the IRB had begun its investigation. It then requested that Dr. Halikas shift his research load to another medical researcher. Eventually, the IRB communicated the conclusions of its investigation to Dr. Halikas and to various agencies and departments within and outside the university. Dr. Halikas then sought to enjoin the University of Minnesota and its IRB from further dissemination of the results of the IRB investigation. He also sought a retraction of information already disseminated, as well as indemnification for attorney fees, and sought redress for violations of due process, as well as for violations of federal research law and contractual breaches.[9]

Judge Rosenbaum of the federal district court took judicial notice under Federal Rule of Evidence 201 of the university's reputation in medical research.[10] He wrote, "Such an institution has a great—and protectable—interest in assuring the integrity and humanity of its research investigations. The IRB is the mechanism by which a medical research institution maintains this integrity and humanity."[11]

Judge Rosenbaum's characterization of the IRB as a norm enforcer signaled that he would give the IRB's determination significant normative weight. He continued:

In this Court's view, an unwarranted intrusion into the IRB's supervisory efforts would inflict grievous harm upon the University and this vital function. Supervision by the IRB protects a vulnerable public. This supervisory power is not without limits. But these limits are delineated in the IRB's own procedures, the law, and federal regulations. On this record, the Court finds that the IRB operated within these procedural limits.[12]

Judge Rosenbaum found that an injunction would impair the IRB's ability to protect the public. Further, he found that the IRB was in sufficient compliance with the procedures in the University of Minnesota General Assurance Agreement and the applicable federal regulations to satisfy the required procedural due process. He indicated that, if the IRB had not operated within the required procedural limits, reasons for following the IRB's guidance would have been weaker. But, having found no reason to override the IRB's determination, or even to balance it against Halikas' claims (because they turned out to be unfounded), Judge Rosenblum dismissed without prejudice Halikas' claim for damages.

It is worth noting that, although questions of moral and cultural difference were not central to the litigation regarding the investigation, they nonetheless gave rise to it. In its initial review of the study, the Minnesota IRB apparently overlooked both cultural and moral differences. This lapse was not a reason, however, for Judge Rosenbaum to override the IRB's investigative judgments. However, a federal judge in New York, addressing another researcher–IRB dispute, was presented with a potentially strong reason to override an IRB's determination.

Marinoff v. *City College of NY*[12] arose on a motion for summary judgment by the City College of New York in a recent (2005) dispute regarding a moratorium on research on philosophical counseling conducted by Prof. Marinoff. City College's IRB had approved Marinoff's proposal to develop the theory and practice of philosophical counseling and to disseminate his research findings.[13] The chair of the IRB had suspended consideration of the renewal of the protocol, however, when he became aware that the College had directed Marinoff to cease his on-campus philosophical counseling pending completion of legal review of the activity. Marinoff was free to submit his research protocol to another IRB within the City University of New York system, and to conduct research with his private clients during the period of the moratorium. The moratorium ended after 20 months. Pursuant to 42 U.S.C. S. 1983, Marinoff sought declaratory and injunctive relief, and compensatory and punitive damages, as well as attorney's fees and costs for alleged violations of his First Amendment rights.

Judge Stein of the federal district court analyzed both Marinoff's interests and conduct and the conduct of the IRB. Marinoff's research, he found, constituted speech on a matter of public concern. Under *Pickering* v. *Board of Educ.*[14] the court attempted to "arrive at a balance between the interest of the teacher, as a citizen, in commenting upon matters of public concern and the interest of the State, as an employer, in promoting the efficiency of the public services it performs through its employees."[15] Judge Stein acknowledged that Marinoff's interest in conducting research was potentially weighty, given that it occurred within the context of an academic setting, but found the actual burden on Marinoff's research to have been light, because the moratorium was limited in time, scope, and degree.

Judge Stein took into account the IRB's concerns and responsibility to assess risks to student-subjects, as well as its concerns regarding institutional liability. Importantly, neither Prof. Marinoff nor Judge Stein intimated that the IRB failed to follow federal research guidelines.

Judge Stein granted the college's motion for summary judgment. Acknowledging that Marinoff's research implicated the First Amendment and involved matters of public concern, Judge Stein nevertheless determined that the reason for supporting the IRB's and college's determinations were stronger. The college had not violated the First Amendment by requiring Marinoff to temporarily cease his on-campus philosophical counseling activities while at City College.

2. Normative Weight Conditioned on Legal Compliance

In both *Halikas* v. *University of Minn.* and *Marinoff* v. *City College of NY*, IRBs had complied with federal regulations. When there is evidence that IRBs have not complied with the regulations that govern them, however, judges are not so open to IRBs' determinations. This was illustrated in *Kus* v. *Sherman Hosp.*, *Gregg* v. *Kane*, and *Grimes* v. *Kennedy Krieger Inst., Inc.* In *Kus* v. *Sherman Hosp.*, an Illinois IRB failed to continue to review a consent form it had approved for experimental intraocular lens surgery. An injured research subject sued both the physician and the hospital for medical battery, as well as for negligence for lack of informed consent and failure to respond adequately to a product recall.

Under the federal guidelines, the hospital was required to conduct "continued review of research." This includes checking to ensure that the form an IRB promulgates is actually being used in the research. Evidence in the record indicated that the consent form approved by the IRB had been modified by the researcher, so that the plaintiff allegedly did not know that he was participating in an experiment.

Despite the IRB's view that its obligations to ensure informed consent had been discharged, the court declined to give the IRB's determination any normative weight. Because there was evidence that the IRB had not complied with the federal regulatory framework,[16] the IRB's determinations forfeited their normative status. In fact, the court found that the medical battery claim could go to the jury.

In another case involving intraocular lens surgery, a Pennsylvania IRB failed to comply with regulations requiring it to make a risk assessment. The court wrote in *Gregg* v. *Kane*:[17]

> Wills' Institutional Review Board (IRB), the body which approved the protocol in which Mrs. Gregg was enrolled, was required by FDA regulations to make an independent risk assessment prior to its approval.... Yet neither the IRB statement granting unconditional approval nor the minutes of IRB meetings discussing the protocol contain any indication that such an assessment was done. Nor is there any other evidence that the IRB performed the independent risk assessment.

The court also noted the IRB had shirked its responsibility as norm enforcer, failing to ensure that research subjects gave informed consent. As a result, not only would the IRB's determination not be given normative weight, but the IRB itself could suffer legal consequences:[18]

> ...as to plaintiffs' claim for lack of informed consent, the FDA regulations make IRBs like the one at Wills responsible for insuring that informed consent will be sought from all prospective subjects, in accordance with fairly detailed standards governing such consent...Given what occurred with respect to informed consent in Mrs. Gregg's case (see *supra*), Wills could be found liable for not fulfilling its FDA-mandated responsibilities.[19]

Although the failure of Will's IRB to assess the risks of the intraocular lens procedure resulted in a loss of summary judgment in *Gregg* v. *Kane*, the jury ultimately decided for the hospital. The results of IRBs' failures can, in some cases, nonetheless, be highly consequential for IRBs. A judge may reason that violation of a federal research regulation can provide evidence of negligence, a rebuttable presumption of negligence, or, in some jurisdictions, constitute negligence *per se*.[20]

That normative weight is conditioned on an IRBs' compliance is also illustrated by *Grimes* v. *Kennedy Krieger Inst., Inc.*,[21] a 2001 decision of the Court of Appeals of Maryland. Mothers of minor research subjects had brought suit against an affiliate of Johns Hopkins University, Kennedy Krieger Research Institute, which had conducted a lead paint abatement study in the 1990s. Kennedy Krieger Research Institute had won summary judgment by Circuit Court for Baltimore City on the basis that there was no contract, no privity, and no special relationship between researchers and subjects that could have been the basis for a researcher's duty to warn the children of risks. In *Grimes*, the Court of Appeals of Maryland vacated that decision and remanded to the District Court.[22]

Judge Cathell harshly criticized that IRB's work. In a discussion of the problems he found in the IRB's review of the research, Judge Cathell included a communication from the IRB chair that suggested noncompliance with federal regulations:

> ...In a letter dated May 11, 1992, the Johns Hopkins University Joint Committee on Clinical Investigation (the IRB for the University), charged with insuring the safety of the subjects and compliance with federal regulations, wrote to Dr. Farfel, the person in charge of the research:
>
> "A number of questions came up.... Please respond to the following points[:]
>
> The next issue has to do with drawing blood from the control population, namely children growing up in modern urban housing. *Federal guidelines are really quite specific regarding using children as controls in projects in which there is no potential benefit* [to the particular children]. To call a subject a normal control is to indicate that there is no real benefit to be received [by the particular children].... So, we think it would be much more acceptable to indicate that the 'control group' is being studied

to determine what exposure outside the home may play
in a total lead exposure; thereby, indicating that these
control individuals are gaining some benefit, namely
learning whether safe housing alone is sufficient to keep
the blood-lead levels in acceptable bounds. We suggest
that you modify…consent form[s]…accordingly."
[Emphasis added.][22]

Judge Cathell thought that this letter demonstrated that the
IRB was willing to help researchers evade federal regulations
designed to protect children.[23] Convinced that the IRB had failed
to do either law or ethics well, he concluded that the circuit
courts had erred in granting summary judgment to Kennedy
Krieger. However, he also went further, using the IRB's work to
represent not only a failure in this case, but also the failure of
IRBs in general.

3. IRB Determinations as Legislative Fact

In the discussion in Chapter 2 of *Spahn* v. *Eisenberg*, we
saw that a judge could use a single HEC's work to represent all
HECs. Judge Cathell did something similar in *Grimes*. He used
the unsatisfactory performance of the Hopkins IRB to represent
IRBs in general, and, ultimately, to provide plaintiffs with protection from research risks outside the IRB system.

Included in Judge Cathell's lengthy critique of the Hopkins
IRB was an assertion that IRBs are primarily in-house organs,
disinclined to make objective assessments. He wrote:

It is clear to this Court that the scientific and medical com-
munities cannot be permitted to assume sole authority to
determine ultimately what is right and appropriate in
respect to research projects involving young children free of
the limitations and consequences of the application of
Maryland law. The Institutional Review Boards, IRBs, are,

primarily, in-house organs. In our view, they are not designed, generally, to be sufficiently objective in the sense that they are as sufficiently concerned with the ethicality of the experiments they review as they are with the success of the experiments.... Here, the IRB, whose primary function was to insure safety and compliance with applicable regulations, encouraged the researchers to misrepresent the purpose of the research in order to bring the study under the label of "therapeutic" and thus under a lower safety standard of regulation. The IRB's purpose was ethically wrong, and its understanding of the experiment's benefit incorrect.[24]

Judge Cathell inferred from the "ethically wrong" purpose and conduct of this particular IRB, and from commentators' work, that "IRBs are, primarily, in-house organs." He used this as legislative fact to justify creating legal conditions under which research subjects could receive greater legal protection.[25]

In each of these cases, the IRB's compliance was probed. In another category of conflicts, it is not the IRB's determinations that are challenged. Instead, plaintiffs are challenging determinations made by health insurance administrators, who, as required by health insurance contracts, rely on IRB determinations.

4. IRB Determinations as Adjudicative Facts

Judges sometimes are asked to resolve disputes between health insurers and an individual (or the individual's survivors) regarding receipt of and payment for arguably experimental treatment. A health insurance contract may include "IRB approval" among the criteria that the plan administrator should consider when determining whether a treatment is covered under the policy. If a research protocol has been approved by an IRB, then, typically, the procedure or treatment given under the protocol is considered experimental, medically unnecessary, or investigational under the contract. If a plaintiff challenges a plan administrator's decisions to refuse reimbursement, judges rely on the language of the contract, including language regarding IRB approval, to help determine whether the plan administrator abused his or her discretion in deciding that an exclusionary clause in the contract applies. Thus, an IRB determination becomes an adjudicative fact relevant to the terms of the contract, and a reason for deciding that a plan administrator's interpretation of the contract was not arbitrary and capricious.

For example, in *Benisek* v. *Rush Prudential HMO, Inc.*, the surviving husband of a woman with a plasma cell disorder filed a complaint under the Employee Retirement Income Security Act claiming that payment for medical services and supplies had been wrongfully denied.[26] Rush Prudential HMO determined that the services and supplies were investigational. Under the plan, a service or supply could be considered experimental or investigational if the HMO determined that the treatment was "reviewed and approved by the treating facility's Institutional Review Board or other body serving a similar function, or if federal law requires such review and approval." The court found that because treatment had been reviewed and approved by the IRB and because the patient signed an IRB-approved consent form, it was reasonable (not "arbitrary and capricious") for the HMO to have determined that the contract language excluded coverage. The defendant's motion for summary judgment was granted. The judge relied on the determination of the IRB to determine what was investigational.

Fuja v. *Benefit Trust Life Ins. Co.* illustrates similar judicial treatment of contractual language in which IRB review triggers exclusion from insurance coverage.[27] Mrs. Fuja's husband sued her insurer for its refusal to pay for high-dose chemotherapy treatment with autologous bone marrow transplantation (HDC/ABMT). He sought injunctive and declaratory relief to prevent the insurer from denying coverage for the treatment under the Employee Retirement Income Security Act. The lower court had entered judgment for the Fujas because the contract's exclusionary language, "in connection with medical or other research," seemed ambiguous. Benefit Trust subsequently paid for the treatment and then appealed the judgment. The appellate court stated that the "most significant" evidence that the HDC/ABMT was provided "in connection with medical or other research" was that the protocol under which Mrs. Fuja had received HDC/ABMT had been approved by an IRB. The court then reversed, holding for the insurance company.

Judges rely on IRB categorizations of studies as research if the fact of IRB review is listed as a condition triggering a contractual exclusionary clause. IRB review becomes an adjudicative fact and a reason to determine that the insurer's decision to exclude treatment from coverage was reasonable.

5. Summary

A degree of judicial receptivity to IRB work results from federal delegation of certain legal–ethical tasks to IRBs. If IRBs complete their tasks in conformity with legal norms, judges use IRB determinations normatively, as was illustrated in Judge Rosenbaum's approach to the University of Minnesota IRB's work in *Halikas* v. *University of Minn.,* and Judge Stein's approach to the City University IRB's work in *Marinoff* v. *City College of NY.*

However, IRB determinations are not always given normative weight. Evidence that an IRB has evaded or failed to comply with relevant research regulations eliminates a key reason for doing so. This problem was illustrated in *Gregg* v. *Kane,* in which there was evidence that the IRB failed to do an independent risk assessment; and in *Grimes* v. *Kennedy Krieger Inst., Inc.,* in which the IRB advised investigators regarding how to recast nontherapeutic research as therapeutic research. In *Grimes,* the noncompliant performance of the IRB also served as legislative fact. Judge Cathell's characterization of IRBs as "in-house organs" was a reason for him to look elsewhere than to IRBs to protect research subjects.

In contract disputes between health insurers and individuals who are insured, IRB determinations can become adjudicative fact and a reason for judges to accept a health insurance administrator's judgment. IRB review can trigger an exclusionary clause in a health insurance contract. This judicial openness to IRB determinations incorporated into contracts was illustrated in *Benisek* v. *Rush Prudential HMO, Inc.* and in *Fuja* v. *Benefit Trust Life Ins. Co.*

Both IRB determinations and HEC determinations have, as their unit of analysis, a rather limited set of facts. HEC determinations typically concern a single patient, whereas IRB determinations typically concern a single study. The type of bioethics communication examined in Chapter 4, the federal bioethics commission report, is not so limited, and presents different questions at the interface of bioethics and law.

Endnotes

[1]42 U.S.C. Sec. 289; 45 CFR Sec. 46.101–46.124.

[2]Wolf SM. Law & bioethics: From values to violence. J Law, Med & Ethics 2004;32:293–301.

[3]Office of Human Research Protections, IRB Guidebook, http://www.hhs.gov/ohrp/irb/irb_guidebook.htm (visited October 25, 2005).

[4]William Head, Appellee, v. John Colloton and Lloyd J. Filer, Appellants, 331 N.W.2d 870 (1983).

[5]Lidz CW, Meisel A, and Roth LH. Mrs. X and the bone marrow transplant. Hastings Cent Rep June 1983;13:17–18. *See also* Davis DS. "Dear Mrs. X...." IRB 1983;5:6–9.

[6]Head v. Colloton at 873.

[7]That law is easily accessed at http://www.hhs.gov/ohrp/humansubjects/guidance/45cfr46.htm (visited October 25, 2005).

[8]James A. Halikas, M.D. v. The University of Minnesota; the Institutional Review Board—Human Subjects Committee, of the University of Minnesota; and its Members Susan A. Barry, M.D., Christopher C. Kuni, M.D., Richard W. Bianco, David R.P. Guay, Pharm.D., Martin L. Gunderson, Ph.D., J.D., Dale E. Hammerschmidt, M.D., and Judith E. Reisman, Ph.D., 856 F. Supp. 1331 (1994).

[9]Halikas asserted violations of federal regulations under 21 C.F.F. 56.108(a), for failure to issue written procedures for hearings; 21 C.F.R. 56.108(c) and 56.110, for conducting an expedited review of the study, 21 C.F.R. 56.113 for suspending all of his human subjects research and failing to give prompt notice and reason for the suspension; and 21 C.F.R. 56.115 for failing to make a record of proceedings Halikas at 1333 footnote 4.

[10]F.R. Evid. 201. "Judicial Notice of Adjudicative Facts"

[11]Halikas v. University of Minn., 856 F. Supp. 1331, 1334 (1994).

[12]Lou Marinoff, Plaintiff, -against- City College of New York; Zeev Dagan, personally and in his capacity as Provost of the City College of New York; John Snyder, personally and in his capacity as Dean for Faculty Relations of the City College of New York; James F. Watts, personally and in his capacity as Acting Dean of Humanities of the City College of New York; Arthur J. Spielman, personally and in his capacity as Chair of the Human Subjects Committee of the Institutional Review Board of the City College of New York; John or Jane Doe, personally and in his/her capacity as officials of the City College of New York., Defendants, 357 F. Supp. 2d 672 (2005).

[13]Marinoff v. City College of N.Y., 357 F. Supp. 2d 672, 678 (2005).

[14]Pickering v. Board of Education of Township High School District 205, Will County, 391 U.S. 563 (1968).

[15]Pickering v. Board of Educ., 391 U.S. 563 at 568 (1968).

[16]Kus v. Sherman Hosp., 268 Ill. App. 3d 771, 775–780 (1995).

[17]Karen J. Gregg and Michael W. Gregg, Plaintiffs, v. Daniel M. Kane, M.D., Stephen L. Trokel, M.D., VISX Inc., and Wills Eye Hospital, Defendants, 1997 U.S. Dist. LEXIS 14269 (1997).

[18]Gregg v. Kane, 1997 U.S. Dist. LEXIS 14269, 11–12 (1997). The evidence as presented was eventually found insufficient to support the claim of negligence. Karen J. Gregg and Michael W. Gregg,

Plaintiffs, v. Daniel M. Kane, M.D., Stephen L. Trokel, M.D., VISX, Inc. and Wills Eye Hospital, Defendants, 1998 U.S. Dist. LEXIS 8437 (1998).

[19]Gregg v. Kane, 1997 U.S. Dist. LEXIS 14269, *12 (1997).

[20]Sherman P. Use of federal statutes in state negligence *per se* actions. Whittier L Rev 1992;13:831, 877–883. *See also* Noah B. Bioethical malpractice: risk and responsibilities in human research. J Health Care L & Pol'y 2004;7:175–241.

[21]Ericka Grimes v. Kennedy Krieger Institute, Inc. Myron Higgins, a minor, etc., et al. v. Kennedy Krieger Institute, Inc., 366 Md. 29, 36 (2001).

[22]Grimes v. Kennedy Krieger Institute, Inc., 366 Md. 29, 39–40 (2001).

[23]Grimes v. Kennedy Krieger Inst., Inc., 366 Md. 29, 38–40 (2001).

[24]Grimes v. Kennedy Krieger Inst., Inc., 366 Md. 29, 44 (2001).

[25]He held that in Maryland a parent, appropriate relative, or other applicable surrogate, cannot consent to the participation of a child or other person under legal disability in nontherapeutic research or other studies in which there is any risk of injury or damage to the health of the subject; that informed consent agreements in nontherapeutic research projects, under certain circumstances, can constitute contracts; and that, under certain circumstances, such research relationships can, as a matter of law, constitute "special relationships" giving rise to duties, out of the breach of which negligence actions may arise; that, normally, such special relationships are created between researchers and the human subjects used by the researchers; that governmental regulations can create duties on the part of researchers towards human subjects out of which "special relationships" can arise; and that the Circuit Court erred in their assessment of the law and of the facts as pled in granting Kennedy Krieger Institute's motions for summary judgment.

[26]Robert Benisek, Independent Administrator of the Estate of Patricia Benisek, deceased, Plaintiff, v. Rush Prudential HMO, Inc., and Rush Prudential Health Plans, 1999 U.S. Dist. LEXIS 10584 (1999).

[27]Kenneth Fuja, as Personal Representative of the Estate of Grace R. Fuja, deceased, Plaintiff–Appellee, v. Benefit Trust Life Insurance Company, Defendant–Appellant, 18 F.3d 1405 (1994).

4   Bioethics Commission Reports

Federal bioethics commissions have existed intermittently since the National Commission for the Protection of Human Subjects of Biomedical and Behavioral Research was created by Congress in 1974. They have included the President's Commission for the Study of Ethical Problems in Medicine and Biomedical and Behavioral Research (1978–1983); The Ethics Advisory Board (1978–1979); the Biomedical Ethical Advisory Committee (1988–1990); the Human Embryo Research Panel (1994); the Advisory Committee on Human Radiation Experiments (1994–1995); the National Bioethics Advisory Commission (1996–2001); and the President's Council on Bioethics (2001–present).

These commissions are created and governed by the Federal Advisory Committee Act, and similar to other such commissions, serve a variety of governmental purposes.[1] They provide information and advice to government; broaden the field from which the government seeks policy advice; legitimize governmental viewpoints; build support for proposals; mask governmental unwillingness to act; deflect criticism from the government for unpopular views; transfer responsibility

From: *Bioethics in Law*
By: B. J. Spielman © Humana Press Inc., Totowa, NJ

for making a decision; give the impression that something is being done; and buy time.[2] Bioethics commissioners have had their own goals for the commissions, which have partially overlapped with those of the federal officials who created them. Among these goals is influencing public policy. Judicial citation of the reports, as well as use by legislatures and government agencies, is viewed as a mark of a bioethics commission's success.[3] Some bioethicists have thought that only a consensus report could be an influential report, but the drive to consensus has, over time, turned out to be not only unnecessary, but potentially inconsistent with federal guidelines for such commissions.[4]

Unlike health care ethics committee (HEC) or institutional review board (IRB) determinations, the material and recommendations in bioethics commission reports are not case- or study-specific; they are topical.[5] The comparatively wide-ranging reports are among the most eclectic of bioethics communications. For bioethicists, attorneys, and judges who are inclined to use them, bioethics commission reports can be a virtual grab bag of material. As a result, for this chapter, unlike the chapters on HEC and IRB determinations, it will be helpful to analyze what judges draw from the reports and for what purposes, as well as how judges treat that material.

1. Reports Provide Adjudicative Facts

The Belmont Report: Ethical Principles and Guidelines for the Protection of Human Subjects of Research was one of the first bioethics commission reports, and has been among the most successful in terms of policy influence.[6] Questions regarding just how receptive judges will be to *The Belmont Report* arise because of its unique history, in particular its unique relation to federal law. Produced by the National Commission for the Protection of Human Subjects of Biomedical and Behavioral Research, it was quickly transformed into federal regulations and became the "charter document for the field of bioethics."[7] *The Belmont Report* has also been incorporated by reference into the contracts, or assurances, into which hospitals enter with the federal government to conduct human subjects research.[8]

The terms of such an assurance were at the heart of a dispute in *Ancheff* v. *Hartford Hosp., et al.*, in which *The Belmont Report* was offered for adjudicative purposes. *Ancheff* v. *Hartford Hosp.* was an evidentiary matter appealed to the Supreme Court of Connecticut in 2002.[9] Mr. Ancheff had brought a medical malpractice action against the Hartford, CT hospital and his health care providers for injuries he suffered after participating in a program involving the drug Gentamicin. He claimed that the defendants had improperly conducted clinical trials and study procedures using Gentamicin; failed to inform him that he was a research participant; failed to obtain his informed consent; and failed to disclose to him the experimental nature of his course of treatment with the drug. The defendants claimed that the hospital protocol under which Mr. Ancheff received Gentamicin was not human subject(s) research.

The hospital had signed an agreement with the federal government setting out the conditions under which human subjects research could occur—an agreement that incorporated *The Belmont Report* by reference. Mr. Ancheff thought the report was evidence relevant to determining the standard to which the defendants should be held—an adjudicative fact. He offered *The Belmont Report* in its entirety as evidence of the line between practice and research. A major section of the report is "Boundaries Between Practice and Research."

Mr. Ancheff had made three different offers of *The Belmont Report* at trial. The trial court had ruled against its admissibility each time. Mr. Ancheff eventually appealed to the Connecticut Supreme Court. The issue on appeal was whether the trial court had abused its discretion in excluding the report from evidence. The basis of the trial court's ruling had been that the probative value of *The Belmont Report* was outweighed by its likely unfair prejudicial effect.[10] In deciding that excluding the report under Rule 403 was not an abuse of discretion, the state Supreme Court wrote:

> As our earlier summary of *The Belmont Report* indicates, it contained a great deal of material that the trial court reasonably could have considered as unfairly prejudicial. First, it purported to be, for the most part, a statement of basic ethical principles, and not to be a statement of the legal standard for securing informed consent. Moreover, it invited the jury, in deciding whether the hospital's Gentamicin program constituted research or medical practice, to think about the Nuremberg War Crimes Trials and the Nuremberg Code, the substance of which the report did not describe, and thus, implicitly, to compare the hospital's conduct with whatever the jurors may have understood those terms to mean. It also invited the jury to engage in a highly abstract and philosophical level of inquiry into such subjects as respect for the autonomy of persons, the notion of self-determination, the concept of beneficence, and the various theories of justice. It invited the jury to think about using children and criminal prisoners as subjects of medical research. It invited the jury to think about the meaning of the physician's Hippocratic Oath, which was neither given in full nor explained in any detail. It invited the jury to compare the hospital's conduct to the infamous Tuskegee study. It invited the jury to compare the hospital's conduct regarding the plaintiff to the complexities of securing informed consent from vulnerable groups such as racial minorities, the economically disadvantaged, the very ill, and the institutionalized. We cannot fault the trial court, as the plaintiff would have it, for determining that submitting this material to the jury would unduly arouse its emotions of prejudice, hostility or sympathy, and would tend to confuse the issues and mislead the jury.[11]

The legal system has never been as receptive to a bioethics commission report as federal lawmakers were to *The Belmont Report* in the late 1970s, 1980s, and early 1990s. The report was, in effect, given quasi-legal status. As a result, Mr. Ancheff may have expected judges to be open to the entire report, and for the purpose he proposed: adjudicative fact.

However, Mr. Ancheff's attempts to have *The Belmont Report* admitted into evidence ultimately failed. The Connecticut court believed that, taken as a whole, the prejudicial effect of *The Belmont Report* outweighed its probative value. We saw in Chapter 1 that ethics testimony in *Rezulin Prods. Liab. Litig.* was vulnerable to exclusion on 403 grounds because the language of the testimony could have displaced legal norms. Here, in *Ancheff*, Rule 403 was a barrier to admission because the historical background for human research protections could be a distraction for the jury. In addition, the Connecticut court believed that *The Belmont Report* could invite the jury "to engage in a highly abstract and philosophical level of inquiry into such subjects as respect for the autonomy of persons, the notion of self-determination, the concept of beneficence, and the various theories of justice." The reports' susceptibility to exclusion from evidence under FRE 403 results from numerous differences between law and bioethics, including the fact that the report's unit of analysis was the topic of human subjects research, whereas the judges' unit of analysis was the case.[12] As a result, this highly regarded bioethics commission report, especially the history and principles contained in it, was seen as a distraction from, rather than as an aid to, the legal task in *Ancheff* v. *Hartford Hosp.*

2. Reports Provide Legislative Facts

A more common use of bioethics commission reports is to provide legislative facts. In the examples that follow, we will see that judges are relatively receptive to what they think to be facts about the world contained in the reports. However, although common, using bioethics commission reports for legislative facts is not straightforward. It can be fraught with difficulties.

One might think that judges would turn to bioethics commission reports for ethical material. However, that is not always the case. In *Grimes* v. *Kennedy Krieger Inst., Inc.*,[13] discussed in Chapter 3, a "prologue" referring to a bioethics commission report opens Judge Cathell's opinion and lays the groundwork to alter the common law of negligence. In the prologue's very first paragraph, the court cites a law review article, which in turn quotes the National Commission for the Protection of Human Subjects of Biomedical and Behavioral Research's report, *Research Involving Children*, for the proposition that consent to research had been virtually unanalyzed by courts and legislatures.[14] The court's statement regarding the status of legal research helped establish that *Grimes* v. *Kennedy Krieger Inst., Inc.* was a case of first impression.[15]

Judge Cathell also used material from another bioethics commission report as legislative fact. He borrowed directly from the National Bioethics Advisory Commission's *Ethical and Policy Issues in Research Involving Human Participants*,[16] stating: "[T]here can be a conflict between the need to test hypotheses and the requirement to respect and protect individuals who participate in research. This conflict and the resulting tension that can arise within the research enterprise suggest a need for guidance and oversight."[17]

These examples of borrowed material are very brief, and, unlike *The Belmont Report* in *Anscheff,* present no evidentiary questions. They do, however, present questions that typically arise if extralegal material is used in judicial opinions: what is being borrowed? Is the way the material is used in the judicial opinion differs from the use in the report? If so, what translation problems arise? Is the judge endorsing one particular theory or approach in the field when there are many competing theories or approaches?[18]

Commission reports, similar to most bioethics resources, mix several modes of inquiry. The material imported into *Grimes* is not the report's distinctively ethical material, even though the report *Research Involving Children* seems highly relevant to the *Grimes* dispute. In a section of his opinion captioned "The Ethical Appropriateness of Research,"[19] Judge Cathell did use ethical materials, such as the World Medical Association's Declaration of Helsinki, but not the ethics in *Research Involving Children*.[20] One reason may be that he disagreed with the ethics of *Research Involving Children*, specifically with its definition of minimal risk[21] and its conclusion that parental consent was adequate for nontherapeutic research on subjects who are very young children.[22]

Judge Cathell did, however, borrow the conclusion of a study of pediatric research law that had been commissioned as background for *Research Involving Children*.[23] The conclusion of the study—that there was no law on point—was useful in establishing that the court "writes on a clean slate."[24] Judge Cathell also borrowed a general policy position from the report. Although the bioethics commission report's policy recommendations focused on what existing IRBs and a proposed federal oversight entity could do in the way of oversight, Judge Cathell, who lacked confidence in IRBs, focused on what courts should do for research subjects. He did not endorse the oversight mechanisms the report preferred but, instead, wanted to limit researchers' "no duty defense" and to create the possibility that injured research subjects would receive redress in the courts.

In summary, *Grimes* v. *Kennedy Krieger Inst., Inc.* does not borrow anything distinctively ethical from the ethics reports. In fact, Judge Cathell's holding contradicts one of the normative positions of *Research Involving Children*: that parental consent to nontherapeutic IRB-reviewed research on children too young to give assent may be morally sufficient.[25] Rather than borrowing ethical reasoning, Judge Cathell borrows legal research and the report's general policy approach. Having examined uses of report materials within the framework Davis presents, we now turn to a distinction that is often relevant in cases in which ethical material is borrowed: representational versus rhetorical use.

3. Representational Versus Rhetorical Uses

Often if nonlegal material (including bioethics material) is imported into judicial opinions, translation problems arise. Scholars disagree regarding how much these translation problems matter, if the material adequately serves a rhetorical purpose in the legal opinion. *Woods* v. *Commonwealth*, discussed in Chapter 2, includes several rhetorical, rather than representational, uses of bioethics commission material. *Woods* v. *Commonwealth* was the 2004 guardian *ad litem*'s appeal of a constitutional challenge to a provision of Kentucky's Living Will Directive Act.[26] The guardian claimed that one of the law's provisions, which permits a surrogate to authorize withholding or withdrawing artificial life-sustaining treatment from a ward or patient who is either in a persistent vegetative state or permanently unconscious, violated liberty interests, was against public policy, and breached "modern ethical standards."

The Kentucky court stated that, if the statute's provision were ultimately found constitutional (which it immediately was), "the issue [would] become how to implement it."[27] Material from the President's Commission Report, *Deciding to Forego Life-Sustaining Treatment*, was used twice to support the court's view of implementation. A section of the opinion captioned "Judicial Oversight" used the following statement, taken from one of the studies included in the commission report, as an epigraph: "Of the approximately 2 million people who die each year, 80% die in hospital and long-term care institutions, and perhaps 70% of those after a decision to forego life-sustaining treatment has been made."[28] The court subsequently reasoned, "Thus it would be logically impossible to require court approval of every decision to withhold of withdraw life-prolonging treatment."[29] The court also used the report's judgment, "decision-making about life-sustaining care is rarely improved by resort to courts"[30] to suggest that judicial oversight was undesirable. Such decisions, the court decided, should most often be made outside the legal system, by HECs.

The Kentucky court borrowed these statements, which seem to represent facts about the world, to delegate norm enforcement to HECs. However, the presentation of these statistics regarding how many people die and where they die, especially in the form of an epigraph, is more consistent with a rhetorical use than a representational use of the material. Also consistent with a rhetorical rather than representational use of the material is a careless inference drawn from the statistics. The statistics were taken from a footnote in *Deciding to Forego Life-Sustaining Treatment*, which, in turn, cites to 1960s and 1970s sources. The court neither placed these 30- to 40-year-old statistics in historical context, nor updated them. More importantly, on the basis of these statistics, the court makes a descriptive inference, "thus it would be logically impossible to require court approval of every decision to withhold or withdraw life-prolonging treatment." The inference is not only not a logical inference, but is flawed, because the statistics are dated and refer to national data on all deaths, whereas the statute at issue would affect only a small fraction of future Kentuckians' deaths.[31]

When the court borrows the statement that "decision-making about life-sustaining care is rarely improved by resort to courts," it reinforces its rhetorical purpose. Despite the empirical appearance of the statement, it was never studied empirically. This kind of statement is what legal scholar David L. Faigman calls "suppositional social science."[32] Its inclusion in *Deciding to Forego Life-Sustaining Treatment* was a rhetorical move to shore up support for health care ethics committees rather than to represent empirical reality, and it serves the same purpose in *Woods v. Commonwealth*.

4. Reports Provide Normative Facts

We noted that Judge Cathell had not used ethical material from the reports he used in *Grimes*. This section explores, in more depth, uses and especially attempted uses of the reports as normative facts. If judges were ever to be open to normative uses of the reports, one would think they would be open to such uses of *The Belmont Report*, because of what I have called its quasi-legal status. Two recent cases tested that hypothesis. *Wright* v. *Fred Hutchinson Cancer Research Ctr.*[33] was a 2002 determination of a defendant's motion for summary judgment before a federal district court in Washington. Plaintiffs were the survivors of several deceased research subjects. The plaintiffs claimed that the research subjects had not been provided with all of the information necessary to understand the risks and benefits of the cancer research protocol in which they were involved, or properly informed of defendants' financial interests. They also claimed that the defendants interfered with the proper workings of the IRB. In addition, they claimed that they were third-party beneficiaries to the contract between the Center and the Department of Health and Human Services, which had been informed by *The Belmont Report*.

Addressing the claim under the caption, "Third Cause of Action: The Belmont Report: Breach of the Assurance greement,"[33] Judge Lasnik noted that, for a third-party beneficiary to have any rights under a contract, he or she must be an intended beneficiary, as opposed to an incidental beneficiary. An incidental beneficiary is a party who stands to benefit from execution of a contract, although that was not the intent of either party. The court decided that the Wright plaintiffs were incidental beneficiaries, who could not enforce the contract, and granted the defendant's motion for summary judgment on this claim. This rather straightforward application of third-party beneficiary law ignored the new Belmont "cause of action" and demonstrates that quasi-legal status is not necessarily a step toward to outright legal status.

We see a similar judicial response in *Abney* v. *Amgen, Inc.*,[34] a 2005 decision by a federal district court in Kentucky. Several victims of Parkinson disease sought a preliminary injunction against the sponsor of a drug study in which they had been participating. The sponsor had terminated clinical trials of the drug, which plaintiffs thought had improved their symptoms. Plaintiffs claimed breach of contract and promissory estoppel. They also claimed that the sponsors breached their fiduciary duty toward them by violating the principles of *The Belmont Report*. Inconsistency with the report, they argued, was a breach of the fiduciary duty created by federal research regulations ("The Common Rule"), because the regulations had been established consistent with the report.[35] Plaintiffs also claimed that *The Belmont Report* created a fiduciary relationship between sponsors of the study and the subjects of the study. Judge Hood found, however, that the Common Rule created no such fiduciary relationship. He found that plaintiffs failed to show a strong likelihood of success on the merits of the fiduciary claim, as well as on their breach of contract claim and promissory estoppel, and declined to grant a preliminary injunction. The plaintiffs' attempt to transform The Belmont Report's quasi-legal status into a legal status equivalent to that of the Common Rule was unsuccessful.

The ambiguity of the status of *The Belmont Report* invites the lawyerly creativity evident in *Ancheff* v. *Hartford Hosp., Wright* v. *Fred Hutchinson Cancer Research Ctr.*, and *Abney* v. *Amgen, Inc.* In the latter cases, proposed uses would have removed the ambiguity, transforming it into clear normative fact. Although the judges rejected these attempts, occasionally a judge does assign legal status, and, thereby, significant normative weight, to a bioethics commission report, even if it lacks the quasi-legal status of *The Belmont Report*. This occurred in *Britell* v. *United States*, a 2002 memorandum and order regarding summary judgment by the US District Court for the District of Massachusetts that addresses the constitutionality of a federal abortion funding ban.[36] The US Supreme Court had previously held that language in the Medicaid statute prohibiting the use of federal funds for abortions except to save the life of the mother was facially constitutional under the Equal Protection Clause.[37] Mrs. Britell, whose husband was in military service, had sought reimbursement for the cost of an abortion of an anencephalic fetus after being denied insurance coverage under the Civilian Health and Medical Program of the Uniformed Service (CHAMPUS). In *Britell* v. *United States*, she alleged that CHAMPUS regulations denying funding for an abortion of an anencephalic fetus were unconstitutional as applied to her because they violated the Equal Protection Clause and advanced no legitimate state interests.

CHAMPUS had refused reimbursement on the grounds that encouraging childbirth is rationally related to the state's interest in potential life and the refusal is rationally related to the state's interest in encouraging women to make the "moral" choice to avoid abortion. Moreover, CHAMPUS argued, attempting to draw lines among categories of disabilities would lead down a slippery slope not warranted by constitutional standards.

In a section of the opinion captioned "The Legal Status of Anencephaly," Judge Gertner used the President's Commission Report *Deciding to Forego Life-Sustaining Treatment*[38] to reason that there was no rational justification for CHAMPUS's refusal to fund Britell's abortion of the anencephalic fetus. Absent a rational justification, funding treatment for ectopic pregnancies and spontaneous abortions but not abortion for anencephaly would violate the Equal Protection clause.

Deciding to Forego Life-Sustaining Treatment had endorsed nontreatment of anencephalic infants. Citing the report's position, Judge Gertner reasoned that because *Deciding to Forego Life-Sustaining Treatment* had indicated too weak an interest in the potential life of anencephalic newborns to require treatment for them, the state could not have a very strong interest in the potential life of anencephalic fetuses. She reasoned that one could locate the legal status of anencephaly at the far end of the spectrum of "potential lives" in which the state had a protective interest.

In addition, Judge Gertner used *Deciding to Forego Life-Sustaining Treatment* to refute an analogy CHAMPUS had drawn between the governmental interest in preserving life asserted in *Wash.* v. *Glucksberg* (one of the assisted suicide cases decided in 1997 by the US Supreme Court)[39] and the governmental interest in preserving the lives of anencephalic fetuses. *Deciding to Forego Life-Sustaining Treatment* had referred to "the limited time the infant could survive after birth." The judge reasoned that, in part because of the brevity of expected life to which the report referred, and because of anencephalic infants' unconsciousness, the potential life interest of an anencephalic fetus was less than that of a terminally ill patient. She also distinguished anencephalic infants from terminally ill patients, who have interests and preferences particular to themselves, such as the circumstances of their deaths.

Finally, to show that CHAMPUS' slippery-slope argument was flawed (and thereby eliminate the last possible rational justification disqualifying this abortion from reimbursement), Judge Gertner stated that *Deciding to Forego Life-Sustaining Treatment* had "endorsed the Baby Doe regulations, and that both the writers of the report and the writers of the regulations had been able to distinguish between anencephaly and other disabilities."[40]

Judge Gertner, thus, used the report's endorsement of non-treatment for anencephalic infants, its statement regarding prognoses for them, and its so-called "endorsement" of the Baby Doe regulations to help interpret the Constitution.[41] Judge Gertner used them to help apply the Equal Protection Clause, showing that there was no meaningful difference between, and, therefore, no rational justification for, reimbursing certain abortion-related medical treatment but not this abortion.

Importantly, Judge Gertner seemed to attribute legal authority to *Deciding to Forego Life-Sustaining Treatment* that was equivalent to that of the Baby Doe regulations, treating them both as official government positions.[42] The plaintiffs in *Abney* v. *Amgen, Inc.* had tried that approach with *The Belmont Report*. Despite its quasi-legal status, they failed. *Deciding to Forego Life-Sustaining Treatment* lacks even the quasi-legal status of *The Belmont Report*. Importantly, on appeal, neither Judge Gertner's decision nor her "normative fact" approach to the report survived.[43]

5. Approaches to Moral Pluralism

Despite the reversal of Judge Gertner's opinion, her treatment of moral pluralism merits attention because of the emphasis that has been placed on consensus-driven reports in the past. In response to CHAMPUS's morals-based justification for the reimbursement restriction, Judge Gertner directly addressed the issue of moral pluralism. She acknowledged the reality that some people believe aborting anencephalic infants is immoral, while clearly recognizing that the plaintiff in the case before her did not.

Judge Gertner was not hostile to morals legislation *per se*. She summarized the law of morals legislation, paying particular attention to what the state must do to make a morals-based argument: The state must articulate a clear moral purpose for the legislation in the text or in the legislative history. If the state were allowed to assert a vague "moral" purpose for legislation, especially in the absence of a clear articulation of that purpose in the text or legislative history, as in this case, then rational review of morals legislation would be meaningless. States could simply claim a "moral" purpose for every law that treated groups differently, Judge Gertner concluded.

Judge Gertner shied away, however, from engaging in morals- or ethics-based reasoning and did not treat *Deciding to Forego Life-Sustaining Treatment* as if it comprised a moral view. She did treat the report normatively, but to do this, presented it as if it were an official government view, despite the fact that this required attributing to the report a legal status it did not have.[44]

We can compare Judge Gertner's approach to moral pluralism to the approach used in *Woods* v. *Commonwealth*, and to that used in *Jeter* v. *Mayo Clinic Ariz.* Mr. Woods' guardian had claimed that, even if the Kentucky statute were constitutional, it violated "modern legal, medical and ethical standards." Like Mrs. Britell, the guardian disagreed with the morality represented by current law, and challenged it. The Kentucky court responded to each part of Mr. Woods' guardian's claim, listing modern legal standards, modern medical standards, and ethical standards supporting the statute.

The Kentucky court listed standards for life-sustaining medical treatment that had been promulgated by the National Center for State Courts as well as standards promulgated by the American Medical Association's Council on Ethical and Judicial Affairs. The majority then proceeded to a five-page description of distinctively ethical standards. It cited Pope John Paul II's 1980 "Declaration on Euthanasia"; Pope Pius XII's 1957 "The Prolongation of Life" and his 1957 "Address to an International Congress of Anesthesiologists, the 1980 Sacred Congregation for the Doctrine of the Faith's Declaration on Euthanasia"; and Joseph Cardinal Bernardin's 1986 book, *The Consistent Ethic of Life*. Based on these standards, the majority concluded the statute did not contravene modern legal, medical, or moral ethical standards.[45]

Dissenting, Justice Wintersheimer characterized the majority opinion as "fatally flawed, seek[ing] moral justification from outdated sources."[46] However, a more important flaw than datedness in terms of the selection of ethical standards is that neither the majority nor Judge Wintersheimer used a representative sample of ethics literature or ethics standards. Why would the judges not use a sample that more accurately reflected the broad spectrum of moral views in their state? One answer is that the rhetorical practices of judges tilt toward institutionalized authorities.[47] The majority hints that it was looking for sources that were recognizably religious and authoritative, when it states, "There is a dearth of written authority on this issue from the viewpoint of religious ethicists. The authority that exists has emanated primarily from sources associated with the Roman Catholic Church."[48] Having thus limited the universe of "ethical standards" to readily accessible and authoritative religious writings,[49] the court could ignore ethical standards from other religious groups or nonreligious sources, such as bioethics commission reports. These omissions occurred despite the fact that *Deciding to Forego Life-Sustaining Treatment*—from which the majority borrowed nonethical material—used a procedural "quality of life" ethic consistent with the majority's own views regarding end-of-life medical treatment decisions.

Strictly speaking, the *Woods* majority did not endorse Roman Catholic morality. The majority's purpose was ostensibly only to show a lack of conflict between the statute and modern ethical standards. Nevertheless, by presenting a one-sided picture of the field (a common problem when judges borrow nonlegal material), the court risked appearing blind to the fact of moral pluralism.

The approach to moral pluralism in *Jeter* v. *Mayo Clinic Ariz.* contrasts both with that in *Britell* v. *United States* and with that in *Woods* v. *Commonwealth*. The Jeters appealed from the dismissal of their lawsuit against the Mayo Clinic Arizona, suing for the alleged negligent destruction or loss of five of the Jeter's frozen pre-implantation embryos, which the clinic agreed to cryopreserve and store.[50] The Jeters argued that the superior court improperly dismissed their complaints because their cryopreserved pre-embryos were "persons" under the Arizona wrongful death statutes. In effect, they were requesting that the wrongful death statute be interpreted as a type of morals legislation.

Judge Kessler of the court of appeals cited *Reproduction and Responsibility*,[51] a report of the President's Council on

Bioethics.[52] No less than a half a dozen times, the report provided a legislative fact for the court. The most important of these legislative facts for our purposes was that describing the tension in the current debate concerning pre-embryos: "Most of the commentators recognize a tension between the relative respect to which embryonic material is entitled and the value of using that material for scientific and medical research."[53] Judge Kessler, having generously cited commentators holding a range of views on this topic, determined, "It is the balancing of these two primary concerns that underscores the need for reasoned legislative, not judicial decision-making as to the nature of a 'person' under the wrongful death statutes. Indeed, it is exactly this balance that led the current President's Commission on Bioethics to recommend that Congress prohibit the use of cryo-preserved pre-embryos in research to those developed beyond ten to fourteen days after fertilization."[54] Explicit acknowledgment of, and familiarity with, a range of moral views led Judge Kessler to affirm the dismissal of a wrongful death claim that he thought would have required endorsing one of those moral views. It also led him to suggest that such moral judgments (in effect, creating morals legislation) were a matter for the legislature. In the *Jeter* opinion, we have frank acknowledgment of moral diversity (unlike *Woods*). We also have an unwillingness to judicially create morals legislation, which might have occurred had the report not clearly highlighted the fact of moral pluralism. Because of the report's approach, Judge Kessler could not have used it as normative fact (as did Judge Gertner with *Deciding to Forego*).

6. Summary

This chapter examines what judges borrow from bioethics commission reports, and how they use it. In the cases presented, judges have borrowed a wide variety of materials from the highly eclectic reports, including legal research (as in *Grimes* v. *Kennedy Krieger Inst., Inc.*), medical prognoses (in *Britell* v. *United States*), suppositional social science (as in *Woods* v. *Commonwealth*), descriptions of the state of an ethical debate (*Jeter* v. *Mayo Clinic Arizona*) and, selectively, policy approaches (as in *Grimes* v. *Kennedy Krieger Inst., Inc.* and *Jeter*). They have used the material for rhetorical as well as representational purposes. The distinctively ethical material in the reports seems not to be adopted openly. Although judges use the report's descriptions of moral diversity, they prefer to use other institution's ethical statements (such as those of the papacy in *Woods* v. *Commonwealth*), to write as if the reports represent official government positions (as in *Britell* v. *United States*), or to ignore the ethics with which they disagree (as in *Grimes* v. *Kennedy Krieger Inst., Inc.*).

In Chapter 5, we examine subpoenaed bioethics scholarship. Our focus is not on judicial use of the material but on factors playing into judicial reasoning regarding it.

Endnotes

[1]Federal Advisory Committee Act, 5 U.S.C. App. 2, 1993, enacted October 2, 1972. Pub. L. no. 92-463, 86 Stat. 770.

[2]Bybee JS. Advising the President: Separation of powers and the Federal Advisory Committee Act. Yale Law J 1994;104:51–128 at 58.

[3]Gray BH. Bioethics commissions: What can we learn from past successes and failures? In: Bulger RE, Bobby EM, Fineberg HV, eds. Society's Choices: Social and Ethical Decision Making in Biomedicine. Washington, D.C.: National Academy Press, 1995: 261–306 at 271. *See* Table 4.

[4]Spielman B. Should consensus be 'The commission method' in the U.S.? The perspective of the Federal Advisory Committee Act, regulations, and case law. Bioethics 2004;17:341–356; Jeter v. Mayo Clinic Ariz., 121 P.3d 1256, 1268 (2005), citing The President's Council on Bioethics. Reproduction and responsibility: The regulation of new biotechnologies, http://www.bioethics. gov/reports/reproductionandresponsibility/index.html (visited

February 7, 2006). There is debate regarding whether consensus should be sought to increase a report's influence if the drive for consensus erases moral differences and "thins" the substance of the reports. *See* Evans JH. Playing God?: Human genetic engineering and the rationalization of public bioethical debate. Chicago: The University of Chicago Press, 2002.

[5]The President's Council on Bioethics report Being human: Readings from the President's Council on Bioethics, http://www.bioethics. gov/bookshelf/ (visited February 3, 2006) is even more eclectic than the unusual bioethics commission report. It contains selections from Homer, Tolstoy, Shakespeare, American folk songs, contemporary fiction and poetry, and even a screenplay.

[6]United States National Commission for the Protection of Human Subjects of Biomedical and Behavioral Research. The Belmont Report: Ethical principles and guidelines for the protection of human subjects of research: appendix. Washington D.C.: Dept. of Health, Education, and Welfare, National Commission for the Protection of Human Subjects of Biomedical and Behavioral Research, 1978. The Belmont Report consists, as its subtitle suggests, of an exposition of ethical principles and guidelines for the protection of human subjects of research. That exposition includes historical references to several research abuses, such as those carried on by Nazis during World War II, and the United States Public Health Service's Tuskegee syphilis study, in which rural black men in the United States were used to study the untreated course of syphilis.

[7]http://www.sfms.org/sfm/sfm205b.htm (visited Oct 11, 2005). Whether the report and regulations are the result of an ethical debate stripped of ethical substance is the subject of Evans JH. Playing God?: Human genetic engineering and the rationalization of public bioethical debate. Chicago: The University of Chicago Press, 2002.

[8]http://www.hhs.gov/ohrp/humansubjects/assurance/filasurt.htm (visited July 26, 2005) Version Date 1/6/2005; Federalwide Assurance (FWA) For The Protection Of Human Subjects; US Department of Health and Human Services (HHS); Office for Human Research Protections (OHRP); A. Terms of the Federalwide Assurance (FWA) for Institutions Within the United

States; 1. Human Subjects Research Must be Guided by Ethical Principles; All of the Institution's human subjects research activities, regardless of whether the research is subject to federal regulations, will be guided by the ethical principles in: (a) The Belmont Report: Ethical Principles and Guidelines for the Protection of Human Subjects of Research of the National Commission for the Protection of Human Subjects of Biomedical and Behavioral Research, or (b) other appropriate ethical standards recognized by federal departments and agencies that have adopted the Federal Policy for the Protection of Human Subjects, known as the Common Rule."

[9]John Ancheff v. Hartford Hospital, et al, 260 Conn. 785 (2002).

[10]Ancheff v. Hartford Hosp., 260 Conn. 785, 804 (2002).

[11]Ancheff v. Hartford Hosp., 260 Conn. 785, 806–807 (2002). Resistance to invoking the Nazi analogy because it is "ethically repugnant" is not unique to the litigation context. Concern has arisen in nonlitigation bioethics contexts as well, for example, in the Human Fetal Tissue Transplantation Research panel, in which the analogy was drawn between subjects of Nazi research and dead fetuses used as a source of materials for research. Childress JF. Consensus in ethics and public policy: The deliberations of the U.S. human fetal tissue transplantation research panel. In: Bayertz K., ed. The Concept of Moral Consensus. Netherlands: Kluwer Academic Publishers, 1994;163–187 at 176.

[12]Jasanoff S. Law's knowledge: Science for justice in legal settings. Am J Public Health 2005;95(S1):S49–58 at S51. Jasanoff points out that the law's knowledge is situated within the four corners of the case. It is conditioned by the principles within which a case arises and has meaning as a cause of legal action.

[13]Ericka Grimes v. Kennedy Krieger Institute, Inc. Myron Higgins, a minor, etc., et al. v. Kennedy Krieger Institute, Inc., 366 Md. 29, 36 (2001).

[14]National Commission for the Protection of Human Subjects of Biomedical and Behavioral Research, Report and Recommendations Research Involving Children. Washington, D.C: DHEW Publication, 1977 at 79–80. The prologue continues, "Our research reveals this statement remains as accurate now as it was in 1977." (Grimes v. Kennedy Krieger Inst., Inc., 366 Md. 29, 36 [2001]).

[15]In the context of the legislative function in this case, the Maryland Court refers several times to the historic research abuses that the Ancheff v. Hartford Hosp. court found too prejudicial to admit as evidence of adjudicative fact.

[16]Ethical and Policy Issues in Research Involving Human Participants. IRB 2001;23:14–16.

[17]Grimes v. Kennedy Krieger Inst., Inc., 366 Md. 29, 102 (2001).

[18]Yovel J, Mertz E. The role of social science in legal decisions. In: Sarat A, editor. The Blackwell Companion to Law and Society. Malden, MA, Oxford UK, and Carlton, Victoria, Australia: Blackwell Publishing Ltd., 2004: 410–431.

[19]Grimes v. Kennedy Krieger Inst., Inc., 366 Md. 29, 41 (2001).

[20]Grimes v. Kennedy Krieger Inst., Inc., 366 Md. 29, 99 (2001).

[21]When the court drew on the concept of "minimal risk" (developed in the report and subsequently adopted in the governing pediatric research regulations) it used it more stringently than either the report or the subsequent regulations had. Kopelman LM. Pediatric research regulations under legal scrutiny: Grimes narrows their interpretation. J L Med & Ethics 2002;30:38–49 at 41. Friedman RL. In defense of the Hopkins lead abatement studies. J L Med & Ethics 2002;30:50–57 at 51.

[22]Grimes v. Kennedy Krieger Inst., Inc., 366 Md. 29, 122–123 (2001).

[23]Annas GJ, Glantz LH, Katz BF. Law of informed consent in human experimentation: Children. In: United States National Commission for the Protection of Human Subjects of Biomedical and Behavioral Research, ed. Report and Recommendations Research Involving Children. Washington, D.C.: DHEW, 1977: 2.1–2.63.

[24]Grimes v. Kennedy Krieger Inst., Inc., 366 Md. 29, 41 (2001).

[25]Research Involving Children United States National Commission for the Protection of Human Subjects of Biomedical and Behavioral Research. Report and Recommendations Research Involving Children. Washington, D.C.: DHEW Publication, 1977.

[26]Matthew Woods, deceased, by and through his Guardian Ad Litem, T. Bruce Simpson, Jr., Appellant v. Commonwealth of Kentucky, Cabinet for Human Resources (Now cabinet for Families and Children), Appellee, 142 S.W. 3d. 24 (2004).

[27]Woods v. Commonwealth, 142 S.W. 3d 24, 28 (2004).

[28]Woods v. Commonwealth, 142 S.W. 3d. 24, 68 (2004), citing President's Commission, Deciding to Forego Life-Sustaining Treatment at 15 n. 1, 17–18 (1983).

[29]Woods v. Commonwealth, 142 S.W. 3d. 24, 69 (2004).

[30]Woods v. Commonwealth, 142 S.W. 3d. 24, 72 (2004) citing President's Commission for the Study of Ethical Problems in Medicine and Biomedical and Behavioral Research. Deciding to Forego Life-Sustaining Treatment: A Report on the Ethical, Medical, and Legal Issues in Treatment Decisions. Washington, D.C.: President's Commission for the Study of Ethical Problems in Medicine and Biomedical and Behavioral Research, 1983 at 247.

[31]The appeal challenges the constitutionality of KRS 311.631, a provision of the Kentucky Living Will Directive Act, insofar as it permits a judicially appointed guardian or other designated surrogate to authorize the withholding or withdrawal of artificial life-prolonging treatment from a ward or patient who is either in a persistent vegetative state or permanently unconscious. Woods v. Commonwealth, 142 S.W. 3d. 24, 28 (2004).

[32]"The term 'suppositional science' refers to two types of 'findings' advanced by social researchers: First, those that on their face are untestable or have not been tested in any fashion whatsoever; and second those that assume the veneer of science (i.e., are forwarded as fully tested propositions) but have yet to be tested adequately." Faigman DL. Normative Constitutional fact-finding: Exploring the empirical component of Constitutional interpretation. U Pa L Rev 1991;139:541–613 at 613 footnote 110.

[33]William Lee Wright, et al., Plaintiffs, v. The Fred Hutchinson Cancer Research Center, et al., Defendants, 269 F. Supp. 2d 1286–1296 (2002).

[34]Edward L. Abney, Barbara Allen, James Day, Robert Green, Delbert Jackson, James Pugh, Roger Thacker and Daniel Hunter Webster v. Amgen, Inc., Defendant, 2005 U.S. Dist LEXIS 14258 (2005). Aff'd, Abney V. Amgen, 2006 U. S. App. LEXIS 7638 6A (2006).

[35]Abney v. Amgen, Inc., 2005 U.S. Dist LEXIS 14258 (2005) at 3 ftnt 11.

[36]Britell v. United States, 204 F.Supp. 2d 182 (2002).

[37]Harris, Secretary of Health and Human Services v. McRae, et al., 448 U.S. 297 (1980).

[38]President's Commission for the Study of Ethical Problems in Medicine and Biomedical and Behavioral Research. Deciding to Forego Life-Sustaining Treatment: A Report on the Ethical, Medical, and Legal Issues in Treatment Decisions. Washington, D.C.: President's Commission for the Study of Ethical Problems in Medicine and Biomedical and Behavioral Research, 1983 at 225–227.

[39]Wash. v. Glucksberg, 521 U.S. 702 (1997).

[40]Britell v. United States, 204 F.Supp. 2d 182, 198 (2002).

[41]Faigman calls this type of legislative fact "constitutional-legislative facts." *See* Faigman DL. Normative Constitutional fact-finding: Exploring the empirical component of Constitutional interpretation. U Pa L Rev 1991;139:541–613 at 553.

[42]Britell v. United States, 204 F.Supp. 2d 182, 191 (2002): b. Legal Status of Anencephaly. Funding the abortion of anencephalic fetuses is not the first occasion on which the federal government, and, indeed, other governmental entities, have had to consider the heartbreaking dilemmas faced by anencephaly. The issue has arisen with respect to the question of whether a doctor has an obligation to provide medical treatment to anencephalics, in the event the woman decides to continue her pregnancy and the fetus is not stillborn. Although not controlling, the right to treatment debate is helpful to this analysis in two respects. First, it offers an analogy to the case at bar—that in other settings anencephaly has been treated as so incompatible with "potential life" that a physician may withhold treatment. Second, it bears on CHAMPUS' concerns that covering abortions of anencephalics will lead down the "slippery slope" to funding abortion of other lethal fetal anomalies, which is inconsistent with the statutory mandate.

[43]Britell v. United States, 372 F.3d 1370–1383 (2004).

[44]Although both Deciding to Forego Life-Sustaining Treatment and the Baby Doe regulations distinguished anencephaly from other conditions, and could, thus, serve one of Judge Gertner's purposes (to refute CHAMPUS' slippery slope arguments), the report didn't "endorse" the regulations. Deciding to Forego Life-Sustaining Treatment was, in fact critical of the regulations.

The report writers thought that "Instead of adding further uncertainty to an already complex situation, the Federal government would do better to encourage hospitals to improve their procedures for overseeing life-and-death decisions." President's Commission for the Study of Ethical Problems in Medicine and Biomedical and Behavioral Research. Deciding to Forego Life-Sustaining Treatment: A Report on the Ethical, Medical, and Legal Issues in Treatment Decisions. Washington, D.C.: President's Commission for the Study of Ethical Problems in Medicine and Biomedical and Behavioral Research, 1983 at 226.

[45]Woods v. Commonwealth, 142 S.W. 3d. 24, 46 (2004).

[46]Judge Wintersheimer himself relied on more recent Roman Catholic statements such as John Paul II's Gospel of Life; Evangelium Vitae (1995); and other recent Roman Catholic texts as well as addresses by Pope John Paul II.

[47]Collier CW. The use and abuse of humanistic theory in law: Reexamining the assumptions of interdisciplinary legal scholarship. Duke L J 1991;41:191–272.

[48]Woods v. Commonwealth, 142 S.W. 3d. 24, 63 (2004).

[49]Briefs were submitted by several Roman Catholic groups.

[50]Belinda L. Jeter, a married woman; William R. Jeter, a married man, Plaintiffs-Appellants, v. Mayo Clinic Arizona d/b/a Mayo Clinic Scottsdale and/or Center for Reproductive Medicine, an Arizona corporation, Defendant–Appellee, 121 P.3d 1256 (2005).

[51]The President's Council on Bioethics. Reproduction and responsibility: The regulation of new biotechnologies, http://www.bioethics.gov/reports/reproductionandresponsibility/index.html (visited February 7, 2006).

[52]Spielman B. Should consensus be 'The commission method' in the U.S.? The perspective of the Federal Advisory Committee Act, regulations, and case law. Bioethics 2004;17:341–356

[53]Jeter v. Mayo Clinic Ariz., 121 P.3d 1256, 1268 (2005), citing The President's Council on Bioethics. Reproduction and responsibility: The regulation of new biotechnologies, http://www.bioethics.gov/reports/reproductionandresponsibility/index.html (visited February 7, 2006) at 123–127 and 223–224.

[54]Jeter v. Mayo Clinic Ariz., 121 P.3d 1256 (2005) at 1268 citing The President's Council on Bioethics. Reproduction and responsibility: The regulation of new biotechnologies, http://www.bioethics.gov/ reports/reproductionandresponsibility/index.html (visited February 7, 2006) at 223.

5 Bioethics Scholarship

1. Introduction

Scholarly work in bioethics often finds its way into judicial opinions. Many of the opinions to which we have referred cite one or more bioethics publications. Ordinarily a judge obtains that work through expert testimony or through a brief, or finds it on his or her own or with the help of a clerk. However, on rare occasions, a subpoena may be used to get a scholar's work into the legal system. If bioethics scholarship, or any other research, seems reasonably calculated to lead to the discovery of admissible evidence,[1] litigants can attempt to compel scholars to testify in court or to produce their research notes. As Robert O'Neil, the distinguished First Amendment scholar writes, however, "scholars and subpoenas coexist uneasily."[2]

Two examples, widely publicized cases in which a bioethicist's work was subject to subpoenas, illustrate that uneasiness. In 2001, Sheldon Zink, a medical anthropologist at the University of Pennsylvania Center for Bioethics, was conducting ethnographic research on the Abiomed Artificial Heart Clinical Trial. She was doing fieldwork at Hahnemann University Hospital in Philadelphia. In November of that year, James Quinn was implanted with an artificial heart. Mr. Quinn and his wife were assigned a patient advocate who was also a physician, Dr. David

From: *Bioethics in Law*
By: B. J. Spielman © Humana Press Inc., Totowa, NJ

Casarett. Dissatisfied with Dr. Casarett's advocacy, Mr. and Mrs. Quinn eventually dismissed him and asked Dr. Zink to assume the role. Dr. Zink resigned her position as an ethnographer (although not her position with the bioethics center), and, independently of the Abiomed program, became the Quinns' advocate until Mr. Quinn's death at the end of August 2002.

Several weeks after Mr. Quinn's death, Mrs. Quinn filed a lawsuit against Dr. Casarett, the maker of the artificial heart, the hospital at which the trial was conducted, and the university affiliate and hospital that employed the majority of his clinical staff. In the lawsuit, Mrs. Quinn complained that no one had given her or her husband a comprehensive picture of what the experiment would entail. Attorneys for the defendants subpoenaed Ms. Zink's correspondence, e-mail, medical records, films, diaries, journals, and interviews.[3] The Quinn's attorney wanted her to give her impressions from the witness stand or by way of deposition.[4]

In response to the subpoenas, Zink began a website and a publicity campaign seeking support and affidavits from fellow anthropologists. Her website warned: "if she does not turn over her field notes she will be forced to stand before a court and then likely jailed for contempt if the court does not rule in her favor."[5] Zink soon became what the Alliance for Human Research Protection characterized as a *cause celebre*."[6] The *Philadelphia Inquirer* reported that Zink was "threatening to go to jail rather than give up the notes."[4] A sociologist who had been jailed in the 1990s for refusing to answer grand jury questions about his research commented, "It's a horrendous position for someone to be put in…but I'm afraid it's also necessary to stand up against courts or against torts."[4]

Ultimately, all of the subpoenas for Zink's work were withdrawn. Zink was not forced to stand before a court; she was not ordered to produce her notes, and was not fined or sent to jail. Eventually the Quinn's case was dismissed.

Another bioethicist's testimony was subpoenaed, in 1996 with at least as much subsequent publicity. Mary Faith Marshall, a bioethicist at the Medical University of South Carolina (MUSC), was subpoenaed by the Center for Reproductive Rights, as well as by MUSC. A suit had been brought against MUSC by women who had been arrested under MUSC's Interagency Policy on Management of Substance Abuse During Pregnancy, whose purpose had been to "ensure appropriate management" of pregnant women who abused illegal drugs. That policy, which had been discontinued in 1994 as part of a settlement with the Civil Rights Division of the US Department of Health and Human Services required, in part, that women who came to MUSC's obstetrical clinic and delivered a child who tested positive for illegal drugs be arrested. The women asserted that the policy violated their rights to keep their medical conditions private, to refuse medical treatment, and to procreate. Dr. Marshall testified, "As the institution's bioethicist, I am of the opinion that the interagency policy fails to meet the institution's norms or standards that have to do with informed consent." Specifically, Marshall explained that "the risk of... arrest and incarceration was not made clear to the patients up front."[7]

Dr. Marshall did not object to the subpoena for her testimony. Rather, it was MUSC's board of directors that objected to her involvement in the case. A delay in her promotion by MUSC resulted in the involvement of the American Association of University Professors. Marshall commented, "Bioethicists, like other researchers have to be free to look at problems objectively and to express their views. If there's a threat of retribution for doing that, bioethics won't be able to survive as a discipline."[8]

Ultimately, neither Marshall nor Zink was compelled to disclose her scholarship. Nevertheless, judges have compelled health care research in the past and surely will have occasion to consider compelling bioethics scholars. In the future, therefore, in this chapter, we consider several hypotheticals as well as the facts in the Zink and Marshall cases. How would a judge reason regarding Zink's case? How would a judge reason regarding Marshall's subpoena, if she had resisted?[9]

Several factors would play into judicial reasoning regarding these matters, including some that are distinctive to bioethics work. The basic framework for that reasoning, however, is set by the rules of procedure. When the case for which the bioethicist's work is sought is a criminal case, the Federal Rules of Criminal Procedure (or a state equivalent) provide the general framework for judicial reasoning regarding subpoenas. If the underlying case is civil, as were the cases in which Zink and Marshall were involved, the Rules of Civil Procedure (or a state equivalent) provide the general framework. Whether civil or criminal, however, judicial reasoning regarding subpoenas is a process of balancing interests within the frameworks the rules provide. The focus in this chapter is on factors that play a role in that balancing process. We look first at factors weighing in favor of compelled disclosure and testimony, and then at factors weighing against disclosure. Next, we examine factors that shape protective orders. Finally, we consider factors judges consider in determining whether to require compensation for the taking of intellectual property. At several points, features distinctive to health care research and to bioethics work in particular, are considered.

2. Compelling the Scholar

2.1. Reasons to Compel

The renowned 20th-century evidence scholar, John Henry Wigmore, thought that anyone who would withhold evidence was not worthy of citizenship, should be expelled from society and required to live as a hermit.[10] Although Wigmore's view may seem extreme today, several factors weigh in favor of compelling testimony or documents. Some apply to any kind of scholarship, and others apply specifically to health research, including bioethics research.

Both private parties and the public have a stake in disclosure of research information.[11] In criminal cases, especially, obtaining accurate information is key to protecting the interests of the public. Imagine that an organ transplant program had been misrepresenting the medical needs of patients to move them up in the queue of potential organ recipients and, thereby, generate more transplant revenues.[12] Information regarding such a program, collected by a bioethicist-researcher, such as Zink, might lead to the discovery of admissible information. Or imagine that an ethics consult has been requested after a complaint by a hospitalized Medicare patient that no physician has spoken to her or examined her for several days. In such situations, federal investigators might demand a great deal of information from health care providers, including, in the first hypothetical, any data the bioethicists may have collected for research on organ transplantation. In the second hypothetical, the demand might include any research data on ethics consultation. The demand for either documentary or testamentary evidence would help protect the public interest in eliminating fraud and reducing health care costs.[13] For these reasons and others, no privilege for bioethics research or any other research has been established in law.[14] The District of Columbia Circuit court in *Burka* v. *United States HHS*[15] observed that, in 1996, confidential research information was "routinely available" in litigation.

If a scholar's privilege were to be recognized in civil cases, it might protect research information in a variety of situations, in addition to those in which Zink found herself. Two hypothetical situations illustrate this point. First, assume that Zink had published her study, and the results of the study were misrepresented by an expert testifying regarding an artificial heart. If litigants wanted to ask Zink herself to testify regarding what her study actually found, she might be subpoenaed to testify regarding the study. Or, second, assume that Zink completed her study, that it was peer reviewed, and that it was published. In subsequent litigation over an Abiomed device, an expert whom one of the parties retains refers to Zink's study, and one of the journal's referees is subpoenaed to testify regarding its strengths and weaknesses.[16] The journal refuses to reveal the reviewers' identities, asserting a need to maintain confidentiality to receive candid, meaningful critiques by referees.

The latter hypothetical roughly parallels the facts in a 1989 case from New York, *Solarex Corp. v. Arco Solar, Inc.*, in which the court declined to endorse a peer review privilege.[17] Arco Solar was a defendant in a patent infringement action. Arco sought to compel the American Physical Society, which published a scholarly journal in the field of physics and which was a non-party to the underlying litigation, to disclose the identity of a scholar who had reviewed a manuscript submitted for publication in the journal. The Society advocated, and Arco opposed, the creation under Rule 501 of the Federal Rules of Evidence of a new, qualified testimonial privilege protecting the identity of reviewers whose critical reviews of manuscripts contribute to the editorial processes of scholarly journals. Although the court did find that interests in maintaining confidentiality outweighed Arco's need for that information, and denied Arco's motion to compel it, the court decided the case on other grounds and refused to endorse the proposed privilege.[18] Even if the scholar ultimately prevails, the lack of a research privilege, along with parties' interests and the public's interests weigh in favor of compelling disclosure.

2.2. Reasons Not to Compel

Although both private parties and the public have interests in disclosure of information, judges balance these interests against reasons supporting nondisclosure. Some of those reasons include that bioethicists may possess information regarding individuals or corporations that is confidential; that ongoing research may be disrupted or invalidated; that prematurely disseminating results of research may prevent publication in a peer-reviewed publication; and that preparing documents for disclosure may be burdensome and expensive. Each of these factors is described below.

One of the distinctive features of bioethics work is that much of it is performed in health care settings, in which confidential information is legally protected in a variety of ways. These protections are not absolute, but tend to facilitate health-related research while providing some protection of confidential information. Judges may have to weigh protections provided by state law as well as by federal law, including certificates of confidentiality,[19] the Health Insurance Portability and Accountability Act (HIPAA),[20] the federal Common Rule,[21] and the US Constitution.

State medical records statutes, registry statutes, and patient's bills of rights provide confidentiality protections that factor into judicial reasoning regarding subpoenas of bioethics material. If a bioethicist's research includes patient information, a judge would consider both state protections and federal protections, including HIPAA.[22]

HIPAA'S Privacy Rule[23] regulates individually identifiable health information, or "protected health information" (PHI).[24] The Privacy Rule applies to three classes of covered entities: health plans, health clearinghouses, and health care providers.[25] A covered entity may not use or disclose PHI, except as permitted or required by the regulations.[26] According to the Privacy Rule, researchers themselves should see only the minimum amount of personal information necessary to conduct the research.[27] How the interactions between state protections and HIPAA affect judicial reasoning regarding subpoenas for patient health information is illustrated in *Northwestern Meml. Hosp.* v. *Ashcroft.*[28] Medical records of women on whom late-term abortions had been performed were subpoenaed for use in a suit in New York challenging the constitutionality of the Partial-Birth Abortion Ban Act of 2003 (discussed in Chapter 1).[29] District court judge Korcoras had held that the production of records after redaction to remove information identifying the patients could be permitted under HIPAA. Nonetheless, he quashed the subpoena because Illinois state law set a more stringent standard for disclosure than does HIPAA.[30] On appeal, the 7th Circuit Court considered whether such a state privilege governed in federal question suits. Writing for the court, Judge Posner found that the HIPAA regulation did not impose state privileges on suits to enforce federal law.[31] The court upheld the district court's decision to quash for several reasons, including invasion of patient privacy and the apparently meager probative value of the records.[32]

Another federal protection for confidential information is the confidentiality certificate. If a bioethicist were conducting research of a particularly sensitive nature, for example research regarding patients receiving mental health or substance abuse services, or treatment for HIV infection, judicial discretion might be constrained by such a certificate. Under the Public Health Service Act, section 301(d), the secretary of Health and Human Services may authorize those involved in biomedical, behavioral, clinical, or other research to withhold identifying characteristics of research subjects from anyone not connected with the research.[33] A confidentiality certificate is the strongest legal protection available; judges cannot compel researchers who have obtained one in any civil, criminal, administrative, legislative, or other proceeding to identify research subjects.[34]

Even when health research information is not so sensitive, if it is individually identifiable, it falls under the Federal Rule for Protection of Human Subjects (the Common Rule).[35] All such research involving human subjects must limit access to private information about the subject unless the investigator has the permission of the subject to do otherwise. The institutional review board (IRB) that approves research protocols must approve the researcher's methods of protecting confidentiality. Even projects that are exempt from full protocol review must include at least an assurance of confidentiality. This assurance of confidentiality may be met in a variety of ways, including substituting codes for identifiers, removing survey cover sheets that contain names and addresses, limiting access to identified data, and storing research records in computers that are password protected and in a locked room (some methods of protecting research data have been recommended specifically to frustrate the possibility of a subpoena—for example, sending the data to a colleague abroad, where it may be analyzed and stored out of reach of subpoena, or keeping the master list of subject names linked to codes abroad).[36]

In some cases of bioethics research involving patient information, judges would also consider the US Constitution. *Ferguson* v. *City of Charleston*,[37] the 2001 appeal to the Supreme Court of the case for which Marshall's testimony was subpoenaed, held that the policy at issue was an unreasonable search in violation of the Federal Constitution's Fourth Amendment. Regarding furnishing patient information to law enforcement authorities, the court wrote: "For purposes of the protection contained in the Federal Constitution's Fourth Amendment against unreasonable searches and seizures, the reasonable expectation of privacy enjoyed by the typical patient undergoing diagnostic tests in a hospital is that the results of those tests will not be shared with nonmedical personnel without the patient's consent."[38] A footnote referred to the Court's previous thinking on the importance of privacy of medical information. In its 1977 decision, *Whalen* v. *Roe*,[39] the Court reviewed a New York state statute that required that records be kept of all prescriptions for controlled substances with the potential for abuse. The *Whalen* v. *Roe* court agreed that there was a constitutionally protected zone of privacy that included the interest in avoiding disclosure of personal matters (although it upheld the challenged law because it adequately protected privacy by limiting access to the lists and building in protections from disclosures).

Confidentiality of research data is important not only to the individuals whose information is at stake, but also to the public, which has an interest in the long-term benefits of research. For some research to yield accurate results, collection of data must not be interrupted or disrupted. The public interest in research is reflected in the provision for waivers of consent under both the Common Rule and HIPAA's Privacy Rule. The consent waivers exist because the Rules' drafters recognized that sometimes research of public importance cannot be conducted if consent is required.[40]

Appreciation for the importance of uninterrupted research has been reflected in judicial reasoning since at least the 1980s. In the subpoena case, *Andrews* v. *Eli Lilly & Co.*, the underlying litigation was a products liability action against drug companies for injuries allegedly caused by *in utero* exposure to the drug, diethylstylbestrol.[41] The defendant-manufacturer subpoenaed information from a cancer registry that was part of a 10-year research project. Federal district Judge Marshall refused to enforce the subpoena request, in part, because he did not want to risk hindering the flow of data into the registry or prematurely disclosing unanalyzed data.[42]

The Federal Rules of Civil Procedure also recognize that a subpoena imposes burdens on the person subpoenaed. If the underlying litigation is a civil case, under Rule 45 of the rules of civil procedure the court must quash or modify the subpoena if it subjects a person to undue burden.[43]

Judge Crabb, whose approach in *Dow Chemical* v. *Allen*[44] demonstrated significant appreciation for academic freedom, nevertheless warned that researchers should not claim merely that the subpoena is "burdensome." Rather, they should provide specifics, so that a judge can properly balance interests. For example, how many subpoenas is the researcher facing? What resources will have to be spent on compliance? Will release of preliminary data endanger the research? Would ongoing research be interrupted if working documents had to be provided or if the researcher were diverted from the work?[45]

The reasoning of judges in criminal cases is, in many ways, differently constrained than the reasoning of judges in civil cases. Under Rule 17 of the rules of criminal procedure, a judge need not quash or modify even a subpoena that is unreasonable or oppressive.[46] Sometimes scholars and journalists have failed to appreciate that a subpoena will be treated differently if the underlying case is criminal rather than civil.[47] For example, Zink's situation, resulting from a civil case, was occasionally analogized to that of Rik Scarce.[48] The underlying investigation in Scarce's case was criminal. He was a sociologist researching radical environmentalism and was jailed for contempt after refusing to cooperate with a federal grand jury investigating a raid on animal experimentation labs at Washington State University that resulted in more than $100,000 damage.

3. Protective Orders

Once the bioethicist who has received a subpoena has made an initial showing of factors weighing against disclosure, the requesting party may demonstrate a need for the information. Judge Crabb lists several factors bearing on the question of the need for the subpoenaed material:[49] the needs of the case; the possibility that the witness is a unique expert or that the discovery could be obtained from another source; the degree to which the discovery sought is necessary to enable the parties to prepare an adequate case or defense; the amount in controversy; the limitations on the parties' resources; the importance of the issues in the litigation; and whether the discovery is cumulative or duplicative.

If this showing is made by the requestor, and the resisting bioethicist moves for a protective order, the court may then consider several types of protections that can accommodate the interests in nondisclosure. More accommodations for confidential information may be available than the subpoenaed bioethicist may realize. One commentator, responding to the Zink case, urged greater caution among bioethicists, writing: "some researchers are promising human subjects a confidentiality they can guarantee only if the researchers are willing to go to jail. Are researchers aware of this risk?"[50] However, jail is not a likely outcome for a bioethics scholar. Under FRCP 26, eight protective options are listed, including an order that identifiers be redacted, and an order that dissemination be strictly limited.[51] Alternatives are not limited to quashing the subpoena or sending a bioethicist to jail, but, as Judge Crabb indicates, include whatever can be developed during negotiation.

4. Compensation

If disclosure or testimony is compelled, judges may consider whether to require the requesting party to compensate the researcher for taking intellectual property. Federal Rule of Civil Procedure 45 attempts to prevent uncompensated taking: "If a subpoena... (ii) requires disclosure of an unretained expert's opinion or information not describing specific events or occurrences in dispute and resulting from the expert's study made not at the request of any party...and if the party in whose behalf the subpoena is issued shows a substantial need for the testimony or material that cannot be otherwise met without undue hardship and assures that the person to whom the subpoena is addressed will be reasonably compensated, the court may order appearance or production only upon specified conditions."[52]

The compensation issue can be complicated by the peculiar nature of bioethics reasoning. A bioethicist may be subpoenaed for many reasons other than that she has researched a problem. The weak and perhaps nonexistent link between research and opinion in bioethics may make it difficult for a bioethicist to receive compensation for loss of intellectual property. Some of the ambiguity regarding the connection between opinion and research is suggested by two of Dr. Marshall's comments. At one point, when academic freedom issues surfaced, Dr. Marshall likened her opinion to that of researchers. However, at another point, she described her opinion as that of "the institution's ethicist." An institutional ethicist's opinion may not "result from" research, but from general knowledge regarding bioethics, skills developed from experience in bioethics, personal values, political or religious ideology, and a host of other factors that would not qualify as intellectual property. A researcher's opinion, by contrast, would be tied closely to research findings.

To be compensable, intellectual property must "not describe specific events or occurrences in dispute" and must "result from the expert's study made not at the request of any party." Arguably, under this rule, neither Zink's research nor Marshall's opinion could have been compensated as intellectual property. Zink was researching the specific events or occurrences in dispute—namely the consent process, in which Mr. Quinn was a participant. If Marshall was studying the policies before the subpoenas were issued, her research might also have described events or occurrences in dispute at MUSC.[53]

An additional complication may arise in bioethics scholarship because compensations for takings of intellectual property are based, in part, on the market for the property. One segment of the market for bioethics publications differs from the markets for scholarly work in most fields. In many fields, forced disclosure of research materials would deprive a scholar of the opportunity have their results published by a peer-reviewed journal as well as of the professional benefits accompanying that achievement. But some bioethics journals trade in what one of the field's leaders called "the crisis *de jour*."[54] Chances for acceptance of scholarship for publication in some journals could actually be enhanced by publicity regarding a bioethicist's ongoing subpoena crisis.[55]

If a subpoena requires disclosure of a corporate trade secret or other confidential research, development, or commercial information, a judge may order production of documents only under specified conditions.[56] Bioethicists who consult for pharmaceutical and biotechnology corporations have expressed concerns both regarding protecting corporate secrets and regarding keeping secret their involvement with corporations. Attempting to address both corporate intellectual property issues, and their own potential exploitation by corporate public relations departments, a joint task force of the American Society of Bioethics and Humanities and the American Society of Law, Medicine & Ethics wrote: "The client might choose to make the entire consultative process confidential as part of its strategic planning process, since even the fact that a company feels it may be confronting an ethical dilemma may reveal a portion of its strategic plan. The ethicists might prefer to keep the process confidential to avoid the appearance of endorsing a particular corporation's effort or product."[57] A judge might consider compensation for a corporation for a taking of intellectual property if a bioethicist's work were subpoenaed and the compelled information exposed the corporation's strategic plan. However, a judge would likely not consider compensation for the bioethicist for exposing the identity of that corporation.[58]

5. Summary

This chapter focused on the factors that judges consider when a bioethicist's work is subpoenaed, rather than offered through briefs or expert testimony or found independently by a clerk or judge. Our main task has been to identify those factors, whether they are distinctive to bioethics work or shared in common with other research. Two features regarding bioethics scholarship are distinctive: it often occurs in settings in which confidentiality protections are already in place;

and the question of compensation is shaped by the nature of, and market for, bioethics.

The *Zink* and *Marshall* cases illustrate both that bioethicists are not immune from subpoenas for their work, and that the distinctive features of bioethics work could factor into judicial thinking regarding subpoenas. Vulnerability to subpoenas may result from bioethicists' proximity to disputes that have their origins in medical settings, clinical research, and corporate decision making. Clinical and corporate bioethicists are more likely than academic bioethicists to have knowledge of disputed events, and, therefore, more likely to be subpoenaed. However, if bioethics work becomes especially vulnerable to subpoenas, it will also, by virtue of the confidentiality protections already in place in health care settings, be well-suited for protective orders. If the bioethics research includes patients' or research subjects' information, judges can be expected to consider fashioning protective orders that take already existing confidentiality protections into account, or to quash the subpoenas, as occurred in *Northwestern Meml. Hosp.* v. *Ashcroft.* Corporate trade secrets will not be treated differently because it is a bioethicist, rather than another type of scholar, who was subpoenaed. If the bioethics research has an intellectual property component, however, and the bioethicist requests compensation for its taking, judges face the complex tasks of determining the relation between research and opinion in bioethics reasoning, and determining economic loss resulting from the taking.

Endnotes

[1]F.R.C.P 26 (b)(1).
[2]O'Neil RM. A researcher's privilege: Does any hope remain? Law Contemp Probl 1996;59:35–49 at 35.
[3]Wilson R. When should a scholar's notes be confidential? An anthropologist involved in a medical lawsuit says she'll go to jail rather than turn hers over. Chron High Educ 2003 May 16;49:A10–12.

[4]Burling S. Researcher fights for her notes. Philadelphia Inquirer, at http://www.philly.com/mld/inquirer/living/health/5301924.htm 03 March 2003 (last visited Mar. 11, 2003).

[5]Brief History at http://www.Freesheldon.org/html/brief_history.html (last visited Dec. 2, 2004).

[6]Alliance for Human Research Protection. Medical confidentiality of research subject in jeopardy—Lawyers subpoena anthropologist. March 9, 2003 at http://www.ahrp.org/infomail/0303/09.php (last visited May 8, 2003).

[7]In 1997, the jury found in favor of the defendants, the sponsors of the interagency policy. The policy, however, was eventually found by the US Supreme Court to violate the women's Fourth Amendment rights against unreasonable searches and seizures. Crystal M. Ferguson, et al., Petitioners v. City of Charleston, et al. 532 U.S. 67 (2001).

[8]Medical University of South Carolina Administration Backs Down. Academe Online 1999;85(4):6 at http://www.aaup.org/publications/Academe/1999/99ja/JA99NB.HTM#sty3 (last visited May 26, 2005).

[9]Matherne JG. Forced disclosure of academic research. Vand L Rev 1984;37:585–620.

[10]Branzburg v. Hayes, 408 U.S. 665, 688 (1972) (quoting 8 John H. Wigmore on Evidence Sec. 2192 at 72) (McNaughton 1961).

[11]F. R. Civ. P. 26b1. Rule 26b1 of the Federal Rules of Civil Procedure, states that "parties may obtain discovery regarding any matter, not privileged, which is relevant to the subject matter involved in the pending action."

[12]Third Chicago Hospital Pays to Settle Transplant Fraud Allegations, Report on Medicare Compliance, No. 42:12 p. 5 November 20, 2003. The US Attorney's Office recently settled a lawsuit with Chicago hospitals alleged to have committed fraud in their liver transplant programs by significantly overstating the seriousness of their patients' conditions.

[13]In re: Subpoena Duces Tecum; United States of America, Plaintiff–Appellee, v. Dwight L. Bailey, M.D.; Family Health Care Associates of Southwest Virginia, PC, Defendants–Appellants, 228 F. 3d 341 (2000).

[14]F. R. Civ. P. 26b1.

[15]Robert A. Burka, Appellant v. United States Department of Health and Human Services; Public Health Service; The National Institutes of Health; National Cancer Institute, Appellees, 318 U.S. App. D.C. 274 (1996).

[16]In 1998, in Cusumano v. Microsoft Corp., however, the 1st circuit court drew an analogy from a First Amendment journalist's privilege to academicians. Microsoft had subpoenaed for use in an antitrust suit the confidential research materials of two university researchers who planned to use them in a book on Netscape. A district court had quashed the subpoenas, reasoning that academic researchers engaged in prepublication research are entitled to the same protection given journalists because discovery of the researchers' confidential work would impede the free flow of information to the public, just as discovery of journalists' work would. Compelled disclosure would have the same "chilling effect" on their speech as if such protections were withdrawn from the press.

[17]Solarex Corporation and RCA Corporation, Plaintiff, v. Arco Solar, Inc., Defendant–Appellant, v. The American Physical Society, Appellee, 870 F.2d 642 (1989).

[18]Solarex Corp. v. Arco Solar, Inc., 870 F.2d 642 (1989). The district court's decision was affirmed on appeal.

[19]42 U.S.C. 241(d).

[20]45 C.F.R. sec. 164.501.

[21]45 C.F.R. Part 46, Subpart A (2004).

[22]HIPAA's Privacy Rule specifically addresses its interaction with state laws in the Preemption of State Law subpart of the Privacy Rule (45 CFR § 160.201 *et seq.*), whereas the preamble to the Privacy Rule provides guidance on the interaction of HIPAA with other federal laws.

[23]Both the Privacy Rule and the Common Rule define research as "a systematic investigation, including research development, testing, and evaluation designed to develop or contribute to generalizable knowledge." 45 C.F.R. sec. 164.501. (Privacy Rule); 45 C.F.R. sec. 102(d)(2004) (Common Rule).

[24]45 C.F.R. § 160.103 (2004).

[25]45 C.F.R. § 160 160.102 (2004).

[26]45 C.F.R. § 164.502(a) (2004).

[27]45 C.F.R. § 164.514(d) (2004).

[28]Northwestern Meml Hospital, Plaintiff–Appellee, v. John Ashcroft, Attorney General of the United States, Defendant–Appellant, 362 F. 3d 923 (2004).

[29]Planned Parenthood Fed'n of Am., Inc., v. Ashcroft, 2004 U.S. Dist. LEXIS 3383 at *7 (2004). In a related California subpoena case, Planned Parenthood Fed'n of Am. v. Ashcroft, US District Judge Hamilton examined patient records in camera to determine whether they were likely to have any probative value. She denied the government's motions to compel production of patient's medical records, reasoning that the government's interest in the marginally relevant patient records did not justify the invasion of privacy and other harms that would result from disclosure. Planned Parenthood Federation of America, Inc., et al., Plaintiffs, v. John Ashcroft, Defendant, 2004 U.S. Dist. LEXIS 3383 (2004).

[30]National Abortion Federation, et al., Plaintiff, vs. John Ashcroft, Defendant. Northwestern Memorial Hospital, Movant, 2004 U.S. Dist. LEXIS 1701.

[31]An HHS regulation authorizes an entity covered under HIPAA, such as Northwestern, to disclose private health information in judicial or administrative proceedings "in response to an order of a court" or in response to a subpoena, discovery request, or other lawful process. 45 C.F.R. sec. 164.512(e)(1)(i).

[32]Northwestern Meml Hosp. v. Ashcroft, 362 F.3d 923, 932–999 (2004).

[33]Guidance on Certificates of Confidentiality, February 25, 2003 Office for Human Research Protections, Department of Health and human Services, at http://www.hhs.gov/ohrp/humsansubjects/guidance/certcon.htm (last visited November 29, 2004).

[34]*See* Gostin LO. Health information privacy. Cornell L Rev 1995; 80:451–528.

[35]45 C.F.R. Part 46, Subpart A (2004).

[36]*See* Office of Research Administration, University of Missouri–St Louis, Confidentiality in Research on Human Subjects, at http://www.umsl.edu/services/ora/compliance/confidentiality.html (last visited May 25, 2005).

[37]Crystal M. Ferguson, et al., Petitioners v. City of Charleston, et al., 532 U.S. 67 (2001).

[38]Ferguson v. City of Charleston, 532 U.S. 67, 78 (2001).

[39]Ferguson v. City of Charleston, 532 U.S. 67, 78 (2001) citing Whalen, Commissioner of Health of New York v. Roe, et al., 429 U.S. 589, 599–600 (1977).

[40]An IRB may permit consent to be waived entirely or altered if: 1) the research involves no more than minimal risk to the subjects; 2) the waiver or alteration will not adversely affect the rights and welfare of the subjects; 3) the research could not be practicably carried out without the waiver or alteration; and 4) whenever appropriate, the subject will be provided with additional pertinent information after participation. 45 C.F.R. 116(d)(2004). An IRB or privacy board may permit use and disclosure of PHI for research without authorization if: 1) there is not more than a minimal risk to the privacy of the individuals, by having a) an adequate plan to protect the identifiers from improper use and disclosure, b) an adequate plan to destroy the identifiers at the earliest opportunity consistent with the conduct of the research, and c) adequate written assurances that the information will not be redisclosed to any other entity except as required by law; 2) the research could not practicably be conducted without the waiver; and 3) the research could not practicably be conducted without access to and use of the information. 45 C.F.R. 164.512(l)(2)(ii)(2004).

[41]Jane Potter Andrews, Plaintiff, v. Eli Lilly & Co., Inc., et al., Defendants; Tracey Ann Taylor, Plaintiff, v. E.R. Squibb & Sons, Inc., et al, Defendants, Paula Renfroe, et al., Plaintiffs, v. Eli Lilly & Company, et al., Defendants; Nancy Deitchman, Plaintiff, v. Rexall Drug Company, et al., Defendants, 97 F.R.D. 494 (1983).

[42]Andrews v. Eli Lilly & Co., 97 F.R.D. 494, 499–500, 502 (1983).

[43]F. R. Civ. P. 45(3)(A)(iv).

[44]The Dow Chemical Company, Intervening Petitioner–Appellant v. Dr. James R. Allen and John Van Miler Respondents–Appellees, and James P. Wachtendonk, et al., Intervening Respondents–Appellees, 672 F.2d 1262 (1982). In 1982, in Dow Chemical v. Allen, the 7th Circuit considered the argument in relation to administrative subpoenas by the EPA of notes, reports, working papers, and raw data relating to animal toxicity studies at the University of Wisconsin. The privilege was not explicitly recognized, but the appellate court concluded that the researcher's interest in academic

freedom had properly figured into the decision regarding whether forced disclosure was reasonable.

[45]Crabb BB. Judicially compelled disclosure of researchers' data: A judge's view. Law and contemporary problems. 1996;59:9–34.

[46]F. R. Crim. P. 17c2.

[47]Coverage in the Chronicle of Higher Education discounted this distinction. Zink's website included the Scarce case, in which the underlying investigation was criminal, along with cases in which the underlying litigation was civil.

[48]Rik Scarce describes his experience in Contempt of Court: A Scholar's Battle for Free Speech from Behind Bars. Walnut Creek, CA: AltaMira Press, 2005.

[49]Crabb BB. Judicially compelled disclosure of researchers' data: A judge's view. Law and Contemporary Problems 1996;59:9–34 at 26–27.

[50]Watson K. Subpoena of Confidential Research: Implications for Informed Consent, ASBH Exchange 2003;6:5, 8 at http://www.asbh.org/resources/exchange/2003 ASBXFal3.pdf (last visited May 31, 2005).

[51]F. R. Civ. P. 26: The court may make an order to protect a party or person from annoyance, embarrassment, oppression, or undue burden or expense, including one or more of the following: (1) that the disclosure or discovery not be had; (2) that the disclosure or discovery may be had only on specified terms and conditions, including a designation of the time or place; (3) that the discovery may be had only by a method of discovery other than that selected by the party seeking discovery; (4) that certain matters not be inquired into, or that the scope of the disclosure or discovery be limited to certain matters; (5) that discovery be conducted with no one present except persons designated by the court; (6) that a deposition, after being sealed, be opened only by order of the court; (7) that a trade secret or other confidential research, development, or commercial information not be revealed or be revealed only in a designated way; and (8) that the parties simultaneously file specified documents or information enclosed in sealed envelopes to be opened as directed by the court.

[52]F. R. Civ. P. 45c3Bii. Notes to the rule specify that the district court's discretion in these matters should be informed by a

number of factors. These factors include "the degree to which the expert is being called because of his knowledge of facts relevant to the case rather than to give opinion testimony; the difference between testifying to a previously formed or expressed opinion and forming a new one; the possibility that, for other reasons, the witness is a unique expert; the extent to which the calling party is able to show the unlikelihood that any comparable witness will willingly testify; and the degree to which the witness is able to show that he has been oppressed by having continually to testify..."

[53]In her testimony, Marshall stated, "As the institution's bioethicist, I am of the opinion that the interagency policy fails to meet the institution's norms or standards that have to do with informed consent." Marshall did publish her research on the MUSC policies subsequent to the subpoena. *See,* for example, Jos PH, Marshall MF, Perlmutter M. The Charleston policy on cocaine use during pregnancy: a cautionary tale. J Law Med Ethics 1995;23:120–128. Nelson LJ, Marshall MF. Ethical and Legal Analysis of Three Coercive Policies Aimed at Substance Abuse by Pregnant Women. Charleston, S.C.: Medical University of South Carolina, Program In Bioethics, 1998. Medical University of South Carolina Administration Backs Down Academe Online 85 (4):6 (1999) at http://www.aaup.org/publications/Academe/1999/99ja/JA99NB.HTM# sty3 (visited May 26, 2005).

[54]Jonathan Moreno states "...[T] the media does provide many of us with a soapbox that should be exploited for purposes other than simply to comment on the ethics crisis *de jour*." Moreno JD. In the wake of Katrina: Has "bioethics" failed? AJOB 2005;5: W18–19.

[55]There is a difference between property which, if used by another, will cause a loss to the owner (rivalrous property), and property which, even if used by another, will cause no loss to the owner. Compelled bioethics material may, in some instances, be nonrivalrous property. *See* Leonard J. Klay, M.D., et al., Plaintiffs, versus All Defendants, Humana, Inc., Humana Insurance Company, Coventry Health Care of Georgia, Inc., f.k.a. Principal Health Care of Georgia, Inc., United Healthcare of Florida, Inc., Health Net, Inc., f.k.a. Foundation Health, et al., Defendants–Appellees, 425 F.3d 977 (2005) at 985.

[56]F. Rule of Civil Procedure 45C3B(i). *See also* Newberg JE, Dunn RL. Keeping secrets in the campus law: law, values, and rules of engagement for industry–university RD part. Am Bus L J 2002;39:187–241.

[57]Brody B, Dubler N, Blustein J, et al. Bioethics consultation in the private sector. Hastings Cent Rep 2002;32:14–20.

[58]F. R. Civ. P. 45C3B(i) states: (B) If a subpoena (i) requires disclosure of a trade secret or other confidential research, development, or commercial information…the court may order appearance or production only upon specified conditions.

6

Reliability of Bioethics Testimony

General Acceptance

1. Bioethics' Eclecticism

Chapter 1 touched on the conditions under which bioethics expert testimony can be helpful in judicial reasoning. Chapters 6–8 examine, in depth, the requirement that such testimony be reliable. Demonstrating the reliability of bioethics expert testimony is potentially a significant barrier to the use of bioethics in law, in part, because bioethics reasoning is so eclectic.

Recall the testimony from the California medical malpractice suit touched on in the Introduction, in which a California man died while waiting for a replacement pacemaker. The three strands of the expert's thinking regarding disclosing physicians' financial incentives were based on at least three different modes of inquiry. One strand was based on the expert's observations of physicians and physician–patient interactions. Another strand was based on literature regarding health care delivery. A third strand was both a prediction

From: *Bioethics in Law*
By: B. J. Spielman © Humana Press Inc., Totowa, NJ

regarding consequences for physician–patient relationships, and an evaluation of those consequences.

Heinrich ex rel. Heinrich v. *Sweet* similarly illustrates that bioethics testimony is produced by a variety of modes of inquiry.[1] The class action suit was brought in 1995 against health care providers who had, decades before, conducted radiation experiments on patients who died shortly thereafter. As summarized in Judge Lynch's opinion, testimony regarding boron neutron capture therapy suggested that:

> ...there was [no] benefit, [or] potential benefit to the individual terminally ill subjects and there was significant evidence of harm, cell death due to necrosis, and therefore it was [the bioethicist's] opinion that in weighing the risks against the benefits that the risks outweighed the benefits and therefore...the research should not be conducted.[2]

The bioethicist's moral judgment—that it was wrong to have conducted the research—depended on several propositions regarding the history of medicine, medical research, and protections for research subject. Those propositions were based on answers to questions such as: What were the review committees' duties at the time the study protocol was submitted to it? What was the state of medical knowledge at that time? What were research standards at the time?[3] A variety of modes of inquiry were required to provide answers to these questions.

2. Reliability of Ethics Strands

Eclecticism in reasoning does not in itself indicate quality, or a lack thereof. However, eclecticism does make it more difficult to assess the reliability of testimony. If the quality of any strand of reasoning offered by an expert is questioned, F.R.Evid. 702[4] or a state equivalent comes into play. The rule was amended in 2000 to incorporate the Daubert trilogy—*Daubert* v. *Merrell*

Dow Pharms.,[5] *GE* v. *Joiner*,[6] and *Kumho Tire Co.* v. *Carmichael*.[7] In that set of decisions, the Supreme Court had determined that F.R.E. 702 required judges to screen scientific and other specialized expert testimony for reliability. Judges could use any of four factors—whether a theory or technique has been tested, whether it has been subjected to peer review and publication, whether there is a high known or potential error rate, and whether it enjoys general acceptance in the relevant community—in assessing the reliability of the methods used in the testimony.[8] These four factors were not meant to be exhaustive. In fact, when the *Kumho* court extended the gatekeeping obligation from scientific to technical and other specialized knowledge, "extensive and specialized experience" was added as a warrant of reliability. These tests were to be used flexibly.[9] Further underlining judicial discretion, the *Joiner* court determined that judicial decisions regarding admissibility would be reviewed on an abuse of discretion standard.[10]

The amendment to F.R.Evid. 702 requires that: "(1) the testimony is based upon sufficient facts or data, (2) the testimony is the product of reliable principles and methods, and (3) the witness has applied the principles and methods reliably to the facts of the case." The Advisory Committee Notes to Rule 702 highlight the importance of assessing each step in the expert's reasoning. 'The amendment specifically provides that the trial court must scrutinize not only the methodology that was used by the expert, but also whether the methodology has been properly applied to the facts of the case.' As the court noted in *In re Paoli R.R. Yard PCB Litig.*, 35 F.3d 717, 745 (3d Cir. 1994): "any step that renders the analysis unreliable… renders the expert's testimony inadmissible. This is true whether the step completely changes a reliable methodology or merely misapplies that methodology."[11]

As illustrated in both *Biddison* v. *Facey* and *Heinrich ex rel. Heinrich* v. *Sweet*, bioethics testimony is produced both by modes of inquiry that are distinctively ethical, and by modes of inquiry that are not ethical. Scrutiny of the methodology of nonethical modes of inquiry should conform to standards of professionals in those fields.[12] *Daubert* v. *Merrell Dow Pharmaceuticals, Inc.* requires the trial court to assure itself that the expert "employs in the courtroom the same level of intellectual rigor that characterizes the practice of an expert in the relevant field."[13] But to what standards should modes of ethical inquiry be held? This question is the focus of the remainder of this chapter, as well as Chapters 7 and 8.[14]

Of the five reliability criteria suggested in *Daubert* v. *Merrell Dow Pharmaceuticals, Inc.* and *Kumho Tire Co.* v. *Carmichael*, two are not useful for the distinctively ethical steps in bioethics reasoning. Distinctively ethical steps are not "testable" and have no "known or potential error rate." Empirical validation ethics testimony can be sought only if descriptive ethics claims are made. Other, more explicitly social warrants, such as "general acceptance" and "peer review," as well as "extensive and specialized experience," must usually suffice for the distinctively ethical strands in bioethics reasoning, if those strands are ever to be considered reliable. The remainder of this chapter explores the general acceptance criterion, or the Frye test, as a warrant of the reliability of the steps in bioethics testimony, whereas Chapters 7 and 8 explore peer review and experience.

3. Steps in Generally Accepted Approaches to Ethical Reasoning

For expert testimony to be demonstrated reliable under the general acceptance test, proponents of the testimony must be able to show that general acceptance of the steps of the reasoning underlying the testimony has been reached.

To begin to demonstrate or assess reliability based on general acceptance or any other criterion, the theory, principle, method, reasoning, or technique that an expert uses must first be identified or described in detail. The next sections focus on the two most prominent styles of distinctively ethical reasoning in bioethics: principlism and casuistry. These modes of reasoning usually comprise at least part of what is distinctively ethical in bioethics testimony. In *Izidor* v. *Knight*,[15] a wrongful death case from Washington, an expert relied partially on principlist reasoning. *In re Baby K*, a Virginia hospital sought a declaratory judgment regarding an anencephalic infant. Bioethicists for each side relied partially on casuistic reasoning.[16]

3.1. Assessing Steps in Principlist Reasoning

Principlism, the most well-known approach to bioethics reasoning, is built on four ethical principles: autonomy, beneficence, nonmaleficence, and justice. Beauchamp and Childress, who developed the principlist approach, characterize the four principles as "only a framework for identifying and reflecting on moral problems."[17] In their view, another step, "specification" is also required before addressing a particular ethical problem. Specification is a process of providing abstract norms with action-guiding content.[18] The first step in establishing the reliability of principlist testimony is, therefore, establishing the reliability of the principles used, and the second is establishing the reliability of the specification.

The following lengthy excerpt from bioethics testimony in *Izidor* v. *Knight* consists of principles and part of a specification. The question at issue was whether Dr. Knight had a duty to revoke a sports authorization form he had completed for Mr. Izidor (a college student) when Dr. Knight learned from a specialist's report that Izidor had been diagnosed with hypertrophic cardiomyopathy. The student died of hypoertrophic cardiomyopathy shortly after playing a few hours of basketball at college.

> Q. Now, tell me about confidentiality as a fundamental principal *[sic]* in medicine, how did that develop and why do we consider it that today?...[19]

Q. Would you consider it a fundamental right of the patient to have that information kept confidential?

A. Yes.

Q. And now you mentioned autonomy. What do you mean by patient autonomy?

A. In earlier times medicine was more paternalistic, by that I mean basically the principles of do no harm and provide a benefit for the patient were prominent features of doctor/patient relationship, it sort of like the old adage the doctor knows best.

Very often in earlier times physicians would prescribe medication, they would recommend a procedure without providing the patient with very much information and providing it in a way that the patient may have felt that he or she did not have a choice because it was sort of a strong benevolent paternalism on the part of the physician, particularly after World War II we have seen the principal [*sic*] of respect for patients arise and in a sense become more dominant over the benevolent paternalism.

So, for example, currently even if a physician has a very important therapeutic medicine or procedure to offer. If the patient, having been fully informed, does not elect to choose that, then the autonomy of the patient must be respected.

Autonomy is the principal [*sic*] on which other doctrines such as informed consent are based and the right to privacy, the right to confidentiality is also a doctrine that grows out of this larger principal [*sic*] of respect for the patient's autonomy.[20]

Q. Now, you mentioned that this is a cardinal principal [*sic*]. What do you mean that this principal [*sic*] is a cardinal principal [*sic*]. By the way, is that still the case [to]day?

A. Yes, I would say when you think about the development of medical ethics in our culture, obviously we're
a culture of great pluralism, there is great ethnic diversity, great religious diversity and one of the questions
in the field of bioethics was related to the question:
Given such pluralism and such diversity, are there any
principles that could be generally accepted by all parties in medical ethics and growing out the Belmont
report—as you know, the president established several
commissions on bioethics to look into various issues
in bioethics, and the Belmont pretty much affirmed
four principles that we refer to as cardinal principles,
that is respect for the autonomy of the patient, the
health care professional to do no harm, to provide a
benefit, and to act justly or fairly with regard to all
patients.

So those four principles are pretty much considered the fundamental principles in medical evidence things.

Q. Where does confidentiality fit in those four?

A. Confidentiality fits under autonomy but also justice or
fairness that is we—in order to respect the autonomy
of the patient, we need to protect the privilege of their
information then under the principal [sic] of justice we
need to apply that to all patients that fit into that same
category.[21]

In this section of testimony, the bioethics expert has begun
to analyze an ethical problem from within a principlist framework. He relies explicitly on the principle of autonomy, which,
despite admittedly "great cultural pluralism and religious diversity," he regards as a principle that "could generally be accepted
by all parties in medical ethics."[22] He suggests that respecting
autonomy requires protecting Mr. Izidor's confidentiality. Both
the principle (autonomy) and the partial specification (protect
confidentiality) track generally accepted and peer reviewed steps

in principlist ethical reasoning. Each could be considered reliable under the general acceptance criterion, although, at this point, the expert has not fully specified exactly when confidentiality must be protected.[23]

3.2. Assessing Steps in Casuist Reasoning

Casuistry, like principlism, is a well-accepted and multi-step mode of ethical reasoning in bioethics. It is much like common law reasoning. The casuist bioethicist reasons analogically from a familiar case to a new case. Although varieties of casuistry are used in bioethics, each version requires at least three steps: identifying the paradigm case, articulating the rule that is associated with the paradigm case, and connecting the paradigm case to the new case by means of a rule. Casuists claim that by comparing a new case to a paradigmatic case in which the rule that should predominate is clear, one can arrive at a sound moral judgment.[24]

The "child of a Jehovah's Witness" case is a paradigm case in bioethics. It presents these circumstances: Jehovah's Witness parents refuse a life-saving blood transfusion for their child on religious or other deeply held grounds; and health care providers consider overriding the parent's decision to save the child's life.[25] The rule is that health care providers may not decide to override parental decisions unless the child's life or health is at stake.

The paradigm "child of a Jehovah's Witness" case was used in casuist testimony by experts on both sides of the *In re Baby K* case. Baby K's mother wanted emergency respiratory treatment for her anencephalic infant; the physicians and hospital thought that doing so was futile. The bioethics expert for the mother used the case to address the question of when the state may override parental medical decision making. The rule by which he connected the paradigm case to the *In re Baby K* case was that overriding parental decision making in health care is permissible only if the parent's decision may result in life-threatening harm to the child. He introduced the paradigm case and rule as he was questioned regarding the best interests of the child.

A. ...[I]t is a mistake to ask the question, What is the one best possible medical intervention for a given patient? But instead we should ask, Is the choice made by the patient or the surrogate for the patient within the range of the tolerable?

Q. In whose judgment?

A. In the judgment of the court or what I refer to here as society's legitimated authority.

Q. Is it fair to say in anyone's judgment but the physician?

A. No, definitely not. I don't think the hospital administrator has any more authority to override on his own than the physician does. Virtually any private citizen— let me restate that. No private citizen would have the right to override the surrogate in such a case. That is a private citizen saying, even though you have decided what you think is in the interest of your child, I am going to intervene unilaterally and override you.

In the cases we are talking about, override you on the basis that we believe your child is better off dead.

That seems to me to be offensive in the worst possible way. If we are going to override legitimate surrogates, that has to be done, it seems to me, through the judicial process and the judge has to have strong reasons why the parent's judgment is intolerable. It is not just that the parent has missed it a little bit, has not quite chosen the best course, but it must be a major offense when it comes to the welfare interest of the child. For example, a life and death situation with a Jehovah's witness. I think we have well established now that is a case where the court's *[sic]* can intervene. But that is not what we are talking about in the present case.[26]

This expert, testifying for the mother, reasoned casuistically from the Jehovah's Witness case using a rule that is generally accepted in the field and political insights regarding the role of officials.[27] The bioethics expert for the hospital used a different rule when responding to a question regarding the mother's desire to continue treatment for the infant:

Q.   …With respect to the religious component of this— this case, in what ethical context should the mother's religious beliefs be considered in coming to a decision as to the ultimate issue in this case? How important are her religious beliefs?

A.   They're important to understand her and her values. They're not overridingly important, not supreme.

Q.   But her—her religious beliefs should be considered by the ultimate decision-maker.

A.   They should be respected and understood, but that— they're not equivalent to—to commands. You can't command—you ought not to be able to command medical treatment to be rendered on the—to a child or withheld when it's valuable and needed for a child, it holds true, in the name of religion. The Jehovah's Witness case that says that parents can die for own beliefs but they shouldn't make martyrs out of their children, I would say that—that the transla- tion in this case, that they shouldn't—so, what's the proper replacement for the martyr? It's—it's that they should accept the inevitability of the death of their children, their child and not attempt to make an impossible ideal.[28]

Even as the hospital's ethics expert testified, he seemed to recognize he was turning the paradigm Jehovah's Witness case on its head. After stating, "The Jehovah's Witness case that says that parents can die for their own beliefs but they shouldn't make martyrs out of their children," he wonders aloud, "what's the proper replacement for the martyr?"

The expert wonders about replacing "martyr" in the standard child of a Jehovah's Witness case because the term suggests a serious harm; but extending the anencephalic child's life, especially when the child never will have experiences of any kind, does not seem like a serious harm in any ordinary sense. What kind of harm is the expert referring to, then? The expert experiments with the idea of "moral harm."

Q. Do you believe that Baby 'K' is experience—is experiencing physical harm from continued ventilator support and medical treatment?

A. From my reading of the Stumpf report and other documents, it's most probably not.

Q. Is it your experience that Baby—your opinion that Baby 'K' is experiencing any harm from the ventilator and other treatments?

A. No. The baby is not experiencing anything in the—in the sense of feeling something. Physiologically her body is reflexing to what is happening, but my understanding is that you need a cerebral cortex to experience anything.

Q. Do you think that there is any harm to maintaining Baby 'K's' life through ventilator support and other medical treatments?

A. Yes.

Q.   And what is that harm?

A.   It's what I call moral harm. And —

Q.   What —

A.   And —

Q.   —what is that?

A.   It's where you're violating a whole lot, in this case, of important moral principles, and mainly violating what society expects of how one ought to behave in this situation. You can cause harm morally without even touching anybody, but in this case, you're touching. You're doing more than touching. You're cutting, putting needles in. A lot of moral harm is being done. That's what's important about this case.

Q.   Do you believe that there is an ethical consensus that moral harm is—is done to anencephalic infants when they are maintained on a ventilator or given other medical treatments?

A.   Yes.[29]

The hospital's expert began with the same paradigm case as did the mother's bioethics expert. He proceeded, however, to invent a rule using a new understanding of "harm": the state or health care providers may override parental medical decision making if significant, though unexperienced "moral harm" would ensue absent their intervention. This bioethics expert's reasoning would fail a general acceptance test, however, because the rule is new. We revisit this testimony in Chapter 7, to examine the peer review criterion, which a judge may also use in reasoning regarding the reliability of this testimony.

4. Summary

This chapter has characterized bioethics reasoning as a composite of several modes of inquiry, and illustrated that eclecticism with expert testimony from *Biddison* v. *Facey* and *Heinrich ex rel. Heinrich* v. *Sweet*. The distinctively ethical strands of reasoning in bioethics testimony can be distinguished from other strands. Key steps in bioethics testimony can be demonstrated as reliable or unreliable using appropriate criteria. These steps can be subjected to the general acceptance test, as illustrated with principlist testimony in *Izidor* v. *Knight* and casuist bioethics testimony from *In re Baby K*.

General acceptance is a social criterion by which judges may assess the reliability of steps in distinctively ethical reasoning. Peer review and publication is another. Chapter 7 explores that criterion, revisiting *In re Baby K* testimony and analyzing additional testimony in a criminal case.

Endnotes

[1]Evelyn Heinrich, on behalf of her husband, George Heinrich, and Henry M. Sienkewicz, on behalf of his mother, Eileen Rose Sienkewicz Jr., Plaintiffs, Appellants, Cross–Appellees, Rosemary Gualtieri, on behalf of her father, Joseph Mayne, and Walter Carl Van Dyke, Representative of the Estate of Walter Carmen Van Dyke, Plaintiffs, Appellants, v. Elizabeth Dutton Sweet and Frederick H. Grein Jr., Representatives of the Estate of William H. Sweet, M.D., and Massachusetts General Hospital, Defendants, Appellees, Cross–Appellants, United States of America, Defendant, Appellee, Estate of Lee Edward Farr, Associated Universities, Inc., and Massachusetts Institute of Technology, Defendants, 308 F.3d 48, 66 (2002 decided), As amended September 16, 2002. US Supreme Court certiorari denied by Heinrich v. Sweet, 2003 U.S. LEXIS 4433 (U.S., June 9, 2003).

[2]Heinrich *ex rel.* Heinrich v. Sweet, 308 F.3d 48, 66 (2002).

[3]Judge Lynch noted that the expert had failed to differentiate between the state of knowledge before the research in question was conducted and the state of knowledge after the research in question had been conducted. That is, the historical strand of reasoning, which provided an underpinning of the ethics reasoning, was flawed. Judge Lynch reasoned that, in part because of that flaw, the jury verdict for the survivors should be vacated. Heinrich *ex rel.* Heinrich v. Sweet, 308 F.3d 48, 66 (2002).

[4]F.R.Evid. 702: Testimony by Experts: If scientific, technical, or other specialized knowledge will assist the trier of fact to understand the evidence or to determine a fact in issue, a witness qualified as an expert by knowledge, skill, experience, training, or education, may testify thereto in the form of an opinion or otherwise, if (1) the testimony is based upon sufficient facts or data, (2) the testimony is the product of reliable principles and methods, and (3) the witness has applied the principles and methods reliably to the facts of the case. As amended effective December 1, 2000.

[5]Daubert v. Merrell Dow Pharms., 509 U.S. 579 (1993).

[6]General Electric Company, et al., Petitioners v. Robert K. Joiner, et al., 522 U.S. 136 (1997).

[7]Kumho Tire Company, Ltd., et al., Petitioners v. Patrick Carmichael, etc., et al., 526 U.S. 137 (1999).

[8]Daubert v. Merrell Dow Pharms., 509 U.S. 579, 597 (1993).

[9]Kumho Tire Co. v. Carmichael, 526 U.S. 137, 156 (1999).

[10]General Electric Company, et al., Petitioners v. Robert K. Joiner, et al., 522 U.S. 136, 142 (1997).

[11]Advisory Committee Note to F.R.E. 702.

[12]Spielman B. Bioethics testimony: Untangling the strands and testing their reliability. J Law, Med & Ethics 2005;33:222–233 at 226–227.

[13]Federal Advisory Committee Note to F.R.E. 702, citing Kumho Tire Co. v. Carmichael, 119 S.Ct. 1167, 1176 (1999).

[14]Imwinkelried EJ. Expert testimony by ethicists: What should be the norm? J Law Med Ethics 2005;33:198–221; Spielman B. Bioethics testimony: Untangling the strands and testing their reliability. J Law Med Ethics 2005;33:222–233; Latham SR. Expert Bioethics Testimony. J Law Med Ethics 2005;33:242–247.

[15]Myrna M. Izidor, as Personal Representative, Appellant, v. Joseph E. Knight, et al., Respondents, 117 Wn. App. 1070 (2003).

[16]*In re* Baby K, 832 F. Supp. 1022 (1993).

[17]Beauchamp TL, Childress JF. Principles of biomedical ethics. 5th ed. New York, NY: Oxford University Press, 2001 at 15.

[18]Beauchamp TL, Childress JF. Principles of biomedical ethics. 5th ed. New York, NY: Oxford University Press, 2001 at 16.

[19]Izidor v. Knight, deposition of Thomas R. McCormick on May 29, 2001, at p. 106 lines 17–19.

[20]Izidor v. Knight, deposition of Thomas R. McCormick on May 29, 2001, at p. 107 line 25 to p. 109 line 6.

[21]Izidor v. Knight, deposition of Thomas R. McCormick on May 29, 2001, at p.109 line 18 to p. 110 line 23.

[22]After additional testimony, the expert concluded that Dr. Knight had no duty to revoke Mr. Izidor's sports authorization, but did have an obligation to protect Mr. Izidor's confidentiality.

[23]Few specifications in bioethics are generally accepted. The variety of principlists' specifications in several "standard" cases is evident in Beauchamp TL. Methods and principles in biomedical ethics. J Med Ethics 2003;29:269–274; Gillon R. Four scenarios. J Med Ethics 2003;29:267–268; Gillon R. *Primum non nocere* in paediatrics. In: Burgio GR, Lantos J, eds. *Primum non nocere* today. Amsterdam: Elsevier Science B.V., 1994:29–38; Macklin R. Applying the four principles. J Med Ethics 2003;29:275–280; Savulescu J. Festschrift edition of the Journal of Medical Ethics in honour of Raanan Gillon. J Med Ethics 2003;29:265–266.

[24]*See* Kuczewski M. Bioethics' consensus on method: Who could ask for anything more? In Nelson HL, ed. Stories and their limits: Narrative approaches to bioethics, New York: Routledge, 1997:134–152 at136.

[25]Gillon R. *Primum non nocere* in paediatrics. In: Burgio GR, Lantos JD, eds. *Primum non nocere* today. Amsterdam: Elsevier Science B.V., 1994:29–38 at 33; Gillon R. Four scenarios. J Med Ethics 2003;29:267–268 at 267.

[26]Fairfax Hospital v. Baby K, deposition of Robert M. Veatch on March 31, 1993, at p. 112 line 8 to p. 113 line 20.

[27]This rule is one of the few that is accepted in bioethics regardless of method. *See* Beauchamp TL. Methods and principles in biomedical ethics. J Med Ethics 2003;29:269–274; Gillon R. Four scenarios. J Med Ethics 2003;29:267–268; Gillon R. *Primum non*

nocere in paediatrics. In: Burgio GR, Lantos JD, eds. *Primum non nocere* today. Amsterdam: Elsevier, 1994:29–38; Macklin R. Applying the four principles. J Med Ethics 2003;29:275–280; Savulescu J. Festschrift edition of the Journal of Medical Ethics in honour of Raanan Gillon. J Med Ethics 2003;29:265–266.

[28]Fairfax Hospital v. Baby K, deposition of John C. Fletcher on April 13, 1993, at p. 143 lines 4–24.

[29]Fairfax Hospital v. Baby K deposition of John C. Fletcher on April 13, 1993, at p. 39 line 3 to p. 40 line 10.

7

Reliability of Bioethics Testimony

Peer Review and Publication

1. Peer Review as a Default Standard

If bioethics testimony is like "strands to a rope," it is important to separate the distinctively ethical strands from other strands, using reliability criteria appropriate for each type of reasoning. In Chapter 6, we looked at two general approaches to distinctively ethical reasoning in bioethics testimony and applied the general acceptance criteria to excerpts of them from *Izidor* v. *Knight* and *In re Baby K*. However, bioethics experts are not methodological purists. Even the distinctively ethical strands of their reasoning are rarely, if ever, purely casuist or principlist. As a result, the noncasuist and nonprinciplist steps must meet some other criteria of reliability. This chapter focuses on peer review, which may serve as a default reliability criterion for steps of bioethics testimony that are not identifiably casuist or principlist. We examine steps in testimony from *In re Baby K* to which the peer review criterion could be applied. In addition, we look at

From: *Bioethics in Law*
By: B. J. Spielman © Humana Press Inc., Totowa, NJ

testimony from a criminal case, *State* v. *Weitzel*,[1] and apply the peer review criterion to it.

Peer review is a process of vetting one's work before other experts so that its flaws may be revealed and its content improved.[2] It has been the "gold standard" of scholarly work for centuries, and has been used by the US judiciary to assess the quality of evidence since the late 1970s.[3] The US Supreme Court considered the significance of a lack of peer review in contested testimony in both *Daubert* v. *Merrell Dow Pharmaceuticals, Inc.* and *Kumho Tire Co.* v. *Carmichael.* In *Daubert*, recalculations of data in previously published studies had not been published or subjected to peer review.[4] The *Kumho Tire Co.* court, disapproving of the fact that a tire failure analyst's four-factor test had not been subjected to peer review or publication, observed "[D]espite the prevalence of tire testing, [no one]…refer[s] to any articles or paper that validate [the expert's] approach."[5] The rationale for this criterion, articulated in *Daubert*, was that "submission to the scrutiny of the scientific community is a component of 'good science,' in part because it increases the likelihood that substantive flaws in methodology will be detected."[6]

1.1. Assessing the Concept of "Moral Harm"

This section illustrates how distinctively ethical steps in bioethics testimony can be subjected to the peer review criterion. Recall from Chapter 6 that the bioethics expert for the hospital in the *Baby K* case argued by analogy from the child of a Jehovah's Witness case. For the analogy to hold, the child must be "harmed" by the parent's religiously based decision. The expert has stated that the anencephalic infant cannot experience anything, but that it may, nevertheless, be harmed. He is questioned by the attorney regarding what kind of "harm" this would be. The bioethicist identifies the harm as a "moral harm," and, when pressed, suggests the concept is part of the peer-reviewed literature of bioethics:

A.   …The baby is not experiencing anything in the—in the sense of feeling something. Physiologically her body is reflexing to what is happening, but my understanding is that you need a cerebral cortex to experience anything.

Q.   Do you think that there is any harm to maintaining Baby 'K's' life through ventilator support and other medical treatments?

A.   Yes.

Q.   And what is that harm?

A.   It's what I call moral harm. And—

Q.   Is there any particular article that you can refer me to that would describe moral harm to anencephalic babies in these circumstances?

A.   No. I can point you to an article about moral harm.

Q.   Okay.

A.   But not to moral harm about—to anencephalic infants under these circumstances.

Q.   What's the article on moral harm?

A.   It's an article by Alistair McIntyre in a book about ethical issues in social and psychological research. He is discussing—

Mr. Coffey: What's the name of the article?

A.   I think it's called "The Harms and Benefits of Research." It's a chapter in a book edited by Ruth Faden, Tom Beauchamp, and a couple of other people. The title of the book is *Ethical Issues in Social and Psychological Research.*

Q.   Have you personally ever written about moral harm?

A.   I think so. I've written about moral suffering. Have I
      written about moral harm? I think so. I know I've
      written about moral suffering.[7]

The concept of "moral harm" is key to this part of the
bioethics expert's reasoning. Because the expert indicates that
the concept had been discussed in the literature, peer review
would likely be proposed as a warrant for the reliability of this
part of the testimony.

But the concept of moral harm as discussed in the
MacIntyre article to which the expert refers is not as the testimony represents it.[8] Writing about three types of harm to subjects of social science research that a social scientist, Donald
Warwick, had discussed,[9] MacIntyre distinguishes moral harm
to subjects of research from harms to their interests, and distinguishes each from moral harm. "Moral harm," he writes, "is
inflicted on someone when some course of action produces in
that person a greater propensity to commit wrongs. Inducing
another to look for the quick and undeserved reward and teaching others to behave in ways that will produce cynicism are
clearly examples of the infliction of moral harm."[10] The expert
has taken the concept out of context. Neither MacIntyre nor
Warwick suggests that extending life could ever create such a
propensity or be a moral harm to an individual, much less to
an individual such as Baby K, who could not experience harms
or interests.

Therefore, although the concept of moral harm is peer
reviewed in bioethics literature, that literature uses the concept in
a way that could not apply to an anencephalic infant. By checking that literature, attorneys would become aware that this part of
the bioethics expert's testimony could not meet the peer review
criterion, and could make that deficiency clear enough to
improve judicial reasoning regarding the testimony.

1.2. Assessing Medical Futility Within a Contract Theory of Ethics

The mother's bioethics expert's testimony in *Baby K* also referred to peer-reviewed work—this time, to the literature on medical futility.

A.   In the debate over what is called futile care, in the literature there is a distinction between physiologically futile care and normatively futile care.

Physiologically futile care is care that is desired by a patient or a surrogate where in the clinician's opinion the intervention will not have the effect desired by the surrogate.

Normatively futile care is care where the clinicians (sic) must concede that the intervention will have the effect desired by the surrogate, but then in their opinion it is normatively of no value.

That second question, whether the care is normatively futile, is a value judgment. And in my view physicians do not have recognized expertise in making value judgments.

I take it in this case there is no dispute that the odds are better that this infant will live if ventilated at least at certain moments in the treatment course than if the infant is not ventilated.

The issue that seems to me to be in contention is the value of preserving that life through the use of the ventilator and presumably other life-sustaining technology as they may be necessary to preserve the life.

I don't see any basis for physicians to claim that they are experts on the religious or philosophical value judgment about the value of vegetative life or unconscious life.[11]

This part of the testimony is not casuist or principlist. It does, however, draw on a distinction that has been widely peer-reviewed in the bioethics literature on medical futility: that between physiological futility and other kinds of futility.[12] The latter, nonphysiological type of futility, which this expert labels "normative futility," describes circumstances in which the expert thinks physicians should have a limited decision-making role. Explaining why that role should be limited, the expert states, "...That...question, whether the care is normatively futile, is a value judgment. And in my view physicians do not have recognized expertise in making value judgments...I don't see any basis for physicians to claim that they are experts on the religious or philosophical value judgment about the value of vegetative life or unconscious life."

The expert's skepticism is also expressed in his published work, which identifies the flaws in generalizations from scientific or technical expertise to expertise in value judgments.[13] The contract theory underlying his critique is also published in a peer-reviewed volume.[14] It is premised on the reality of moral pluralism—particularly the differences in values between physicians as a group and the public—and examines boundaries that should be observed to respect those moral differences. That difference was central to the conflict in this case, in which the infant's mother had claimed that health care providers' actions violated her First Amendment right to expression of religion.

Having rejected the notion that physicians should make life-and-death value judgments for patients, this bioethics expert concludes that physicians are obligated to treat the infant. The obligation, according to one of the expert's publications to which the attorney will refer, arises as a "condition of the monopoly privileges of licensure:"

> A.   ...Insofar as the surrogate is saying, I believe life is
>   precious and I should have the right to have my daugh-
>   ter's life preserved as long as medicine can provide
>   that service, that is precisely what I am saying.

It is unfortunate that there may be these temporary situations where no physician is willing to provide a life-sustaining intervention when a surrogate in good faith wants the child's life saved.

But the alternative is for the physician to be empowered to say, I can figure out that this child is better off dead.

Q. No. Isn't there another alternative, Dr. Veatch, and that alternative is simply, I won't do it?

A. But saying, I won't do it, when no one else will, constitutes abandonment of the patient.

Q. On page I think it is fifteen you write, we find a limited duty to treat with life-prolonging technology as a condition of the monopoly privileges of licensure.

Do you see that?

Ms. Palmer: Is it under sub-part two?

The Witness: Page fifteen, the second paragraph, is it? You are in the middle of a sentence?

Ms. Palmer: Yes. Yes, you are.

By Ms. Krebs:

Q. We find it a limited duty to treat with life-prolonging technology as a condition of the monopoly privileges of licensure.

As the attorney scrutinizes the expert's article, she questions the extension of the duty the expert has written about to the case at hand.

And you say, this duty is limited to cases where patients insist on the intervention while competent or the clinician is in an ongoing clinical relationship and where no one else is willing.

Does this rule apply to the *Baby K* case because she was
never competent?

A.   Thank you for pointing that out. I hadn't noticed that.
     What was intended here applied to cases in which the
     patient or surrogate insists on intervention while
     competent.

Q.   I am sorry. I have to go back. Hold on.

A.   Certainly it is my intention that this apply to legiti-
     mate surrogate decision making if the surrogate is
     competent.[15]

Note that, when questioned, the expert acknowledges that
he is using the rule differently in the testimony than in his peer-
reviewed work. The literature referred to patients' requests for
treatment, rather than requests of their surrogates. The expert's
theory, the concept of generalization of expertise, and the distinction between physiological futility and other futility used in the
testimony could each meet a peer review criterion. Judges might
have different opinions, however, regarding whether the expert's
extension of the rule from patients to surrogates is significant for
reliability purposes.

1.3. Assessing the Rule of Double Effect

A key step in testimony from *State* v. *Weitzel* can also be tested
by the peer review criterion.[16] Dr. Weitzel, a psychiatrist, had been
tried on four counts of murder for prescribing opiates to terminally
ill elderly patients. An ethics expert testified for the psychiatrist.

The rule of double effect, which is historically religious, is
summarized in Beauchamp and Childress' classic *Principles of
Medical Ethics*. Most formulations of the rule of double effect
identify four conditions or elements that must be satisfied for
an act with a double effect to be justified. Each is a necessary

condition, and together they form sufficient conditions of morally permissible action:

1. The nature of the act. The act must be good, or at least morally neutral (independent of its consequences).
2. The agent's intention. The agent intends only the good effect. The bad effect can be foreseen, tolerated, and permitted, but it must not be intended.
3. The distinction between means and effects. The bad effect must not be a means to the good effect. If the good effect were the direct causal result of the bad effect, the agent would intend the bad effect in pursuit of the good effect.
4. Proportionality between the good effect and the bad effect. The good effect must outweigh the bad effect. That is, the bad effect is permissible only if a proportionate reason compensates for permitting the foreseen bad effect.[17]

The attorney questions the ethics expert in *Weitzel* regarding the double effect reasoning he has used:

Q.   You testified before about the doctrine of double effect. Do you remember that?

A.   Yes.

Q.   And does that have application to the circumstances that you reviewed in this case?

A.   I believe it does, yes.

Q.   And tell us why.

A.   I think it meets most of, if not—some in part and some—in some cases all of the four requisites for that doctrine.

Q. And why don't you explain please if you could what the requisites are and how it is relevant to what you determined based upon your evaluation.

A. Well, the first is that there does have to be a—an intent to treat and that the treatment has to have—be justified by the clinical circumstances. That there is a foreseeable harm that can come from the treatment. Otherwise, there would be no—no double effect possible, that— that—…Thirdly, that the ends cannot justify the means. That is to say, you cannot palliate by causing. In this case, for instance, you cannot palliate by causing death, but death can be a foreseeable occurrence of palliation. That is to say, you know, you can 100 percent guarantee relieving pain by killing people. Dead people feel no pain, presumably. But that can't—you can't purposefully induce the outcome you want by causing the unwanted effect.

That's—and the last and fourth is that there must be some reasonable attempt to—to garner informed consent. And in these particular cases, patients could not give informed consent, and so informed consent would have to be by proxy, either by legal proxy or by associated proxy, family members or a patient's advanced directives or some—something similar to that nature…

By Mr. Stirba:

Q. Does the fact or the concept of hastening one's death through medication, does that have any relevance to the concept or the doctrine of double effect?

A. Well, the doctrine of double effect exists almost universally throughout every medical treatment. And the more dangerous medical treatments are, the more it applies. Most recently, it's gained attention and has been on the forefront of issues including legislative ones we talked about last time because of this issue of—of the—the necessity to treat pain and to mitigate suffering and the concern and fears that physicians have of being prosecuted if death occurs, so the ethical standard or the ethical doctrine of double effect has been sort of reinvoked and looms large specifically to address that.

Q. Is that doctrine well recognized in the field of medical ethics for the practice of medicine?

A. Yes.[18]

The rule of double effect is well-recognized in the field of bioethics, as the expert testified. In fact, each of the four steps Beauchamp and Childress describe has been extensively peer-reviewed in bioethics and philosophical as well as theological ethics. The expert has, however, substituted informed consent for one of the four standard double effect conditions. In doing so, he has created a new list of four conditions. And when asked whether double effect reasoning applies to the circumstances of the four deaths, the expert replied "Yes" but then qualifies his answer, adding, "it meets most of, if not—some in part and some—in some cases all of the four requisites for that doctrine." If this were a situation in which a demonstration of reliability was required, peer review could, in theory, provide a warrant for the reliability of the rule of double effect and each of its steps. However, whether a judge would find that this testimony actually met the criterion is open to question.

2. Limits of Peer Review

Peer review and publication is the gold standard of scholarly work. Paradoxically, it may serve as a default reliability criterion because so little in bioethics testimony is methodologically pure enough to be casuistry or principlism. For those reasons, it is important to discuss the limits of peer review, and the limits of bioethics testimony in relation to it. If peer review and publication are offered as a warrant for the reliability of bioethics testimony, judges unfamiliar with the field may consider several questions regarding the significance and limits of bioethics peer review. Criticisms of peer review in scientific research and social science research are common, and may apply to bioethics peer review as well. Criticisms include that the level of agreement between peer reviewers for journals is no higher than what might be that expected by random chance;[19] that the process screens out articles based on the reviewer's attitudinal predispositions rather than methodological validity;[20] that reviewers give article submissions relatively little scrutiny before judging them;[21] that the peer review process contains a bias for established scholars from prestigious institutions;[22] that the assessments of reviewers are influenced by everything from typeface used in submissions to conflicts of interest of the peer reviewers;[23] that conferences are sometimes open door invitations to present as long as presenters pay the registration fee;[24] that workshops may be opportunities to disseminate practice strategies and not to provide a critical forum for peer review;[24] and that conferences allow speakers to present only summaries of their ideas and ignore the audiences' comments.[24]

Little is known regarding what occurs in peer review processes in bioethics. Criticisms of peer review are not unheard of in the field. One bioethicist has observed:

> [B]ad publication ethics are not unknown in bioethics.... Authors may worry that some work is rejected by a journal because it takes a theoretical line disfavoured by a reviewed or editor. Multiple publication of work is reasonably common. Most frequent of all is publication work, especially in medical or nursing journals, entirely lacking in rigour or originality, simply because of who the author is, or because it is the first article on 'informed consent in Xology' (even though it is the 954th article on the basics of the law and

ethics of informed consent). This latter is a venial sin, but when it permits the author to claim some 'expertise' in ethics, or to tip the scales of his bid for tenure, we might feel that it is dishonest.[25]

Some unevenness can be expected in peer review of bioethics work because of the peculiar nature of the field. First, because bioethics is multidisciplinary, diffusion of knowledge (and knowledge about the work that other scholars are doing) is haphazard. It may be difficult for an editor to determine who is a "peer" capable of judging a particular work in bioethics.[26] The pool of those who are expert in certain aspects of bioethics problems vary in training, experience, types of employment, and visibility to editors. The editor's task of identifying peers to review multidisciplinary work may be quite difficult.

Second, the problem of finding the right reviewers often requires locating not only competent reviewers from within several disciplines or fields but also reviewers who are competent to review the links the author makes between fields. There is often confusion and profound disagreement regarding what standards should be in interdisciplinary and multidisciplinary work.[27]

Third, if what is being reviewed is not science, questions arise regarding whether peer review is a warrant for reliability, or for something else. Commenting on peer review problems in the field of clinical psychology, Shuman warned:

> In science, peer review suggests a level of agreement on the validity of the science, but what does it mean with respect to clinical writing? Peer review may mean it presents an interesting idea, but not necessarily a valid one. It may mean that the ideas agree with those of the editorial consulting reviews and the editor, which at best is a judgment about the writings'...consistency with the views of other practitioners.[28]

Shuman's question regarding peer review in clinical psychology is salient for bioethics peer review. Does publication mean only that the work presents an interesting idea, and not necessarily that it is based on a sound approach? This was the complaint of one prominent medical ethicist. Jonsen observed that a lack of close criticism of the literature in bioethics sometimes results in acceptance and even "canonization" of literature that is provocative, but not necessarily sound.[29] Those attempting to apply a peer review criterion may, following Jonsen, scrutinize which material from bioethics publications and conferences has been appropriately and thoroughly vetted and which has not.

3. Summary

This chapter has examined ethics reasoning in testimony from *In re Baby K* and *State* v. *Weitzel*, showing how peer review and publication could be used as a warrant for reliability. It has also offered a critique of this "gold standard," so that those attempting to use it will better understand its limits. Finally, it has illustrated that bioethics is methodologically eclectic even within its ethical strands.

Endnotes

[1]http://www.weitzelcharts.com (last visited January 27, 2006). Weitzel v. Div. of Occupational & Professional Licensing of the Dept. of Commerce of the State of Utah, 240 F.2d 871 (2001) was brought after the criminal case was dropped. Psychiatrist 'innocent' in 5 Utah deaths, United Press International, Salt Lake City, November 23, 2002.

[2]Sales BD, Shuman DW. Experts in court: Reconciling law, science, and professional knowledge. Washington, D.C.: American Psychological Association, 2005 at 58.

[3]*See* National Commission on Egg Nutrition and Richard Weiner, Inc., Petitioners–Appellants, v. Federal Trade Commission, Respondent–Appellee, 570 F.2d 157 (1977); AFL-CIO v. Marshall, 199 U.S. App. D.C 54 (1979); Peter H. Forsham, et al.,

Appellants, v. Joseph A. Califano, Secretary of the Department of Health, Education and Welfare, et al., 190 U.S. App. D.C. 231 (1978).

[4]Daubert v. Merrell Dow Pharmaceuticals, 727 F. Supp. 570, 575 (1989).

[5]Kumho Tire Co. v. Carmichael, 526 U.S. 137, 157 (1999).

[6]Daubert v. Merrell Dow Pharms., 509 U.S. 579, 593–594 (1993).

[7]Fairfax Hospital v. Baby K, deposition of John C. Fletcher on April 13, 1993 at p. 39 line 3 to p. 40 line 5.

[8]Joel Feinberg's Harm to Others, which describes the concept of moral harm as central to the teaching of Socrates, Plato, and the Stoics had been published several years before. However, Feinberg understood the moral harm discussion to be regarding whether a morally degraded character is itself a harm independent of its effect on its possessors' interests. Feinberg J. The moral limits of the criminal law: harm to others. New York: Oxford University Press, 1984–1988.

[9]Warwick DP. Types of harm in social research. In: Beauchamp TL, Faden RR, Wallace RJ Jr., Walters L, eds. Ethical issues in social science research. Baltimore and London: Johns Hopkins University Press 1982:101–124.

[10]MacIntyre A. Risk, harm, and benefit assessments as instruments of moral evaluation. In: Beauchamp TL, Faden RR, Wallace RJ Jr., Walters L, eds. Ethical issues in social science research. Baltimore and London: Johns Hopkins University Press 1982: 175–189 at 178.

[11]Fairfax Hospital v. Baby K, deposition of John C. Fletcher on April 13, 1993 at p. 49 line 9 to p. 51 line 7.

[12]*See* Veatch RM, Spicer CM. Medically futile care: The role of the physician in setting limits. Am J L & Med 1992;18:15–36; Youngner SJ. Who defines futility? JAMA 1988;260:2094–2095 at 2094; Guidelines on the termination of life-sustaining treatment and the care of the dying: a report of the Hastings Center. Briarcliff Manor, NY: Hastings Center; 1987 at 32.

[13]Most notably Veatch RM, Spicer CM. Medically futile care; The role of the physician in setting limits. Am J L Med 1992;18:15–36.

[14]Veatch RM. A theory of medical ethics. New York: Basic Books, 1981.

[15]Fairfax Hospital v. Baby K, deposition of Robert M. Veatch on March 31, 1993 at p. 146 line 10 to p. 149.

[16]Weitzel v. Div. of Occupational & Professional Licensing of the Dept. of Commerce of the State of Utah.

[17]Beauchamp TL, Childress JF. Principles of biomedical ethics. 5th ed. New York, NY: Oxford University Press, 2001 at 160, ftnt. 37.

[18]Weitzel v. Div. of Occupational & Professional Licensing of the Dept. of Commerce of the State of Utah, 240 F.2d 871 (2001).

[19]*See* Rothwell P, Martyn CN. Reproducibility of peer review in clinical neuroscience. Brain 2000;123:1964–1969. *See also* Cole S, Cole JR, and Simon GA. Chance and consensus in peer review. Science 1981;214:881–886.

[20]*See*, for example, Ceci SJ, Peters D, Plotkin J. Human-subjects review, personal values, and the regulation of social–science research. Am Psych 1985;40:994–1002 at 1001; Horrobin DF. The philosophical basis of peer review and the suppression of innovation. JAMA 1990;263:1438–1441.

[21]*See* Chan EJ. The "brave new world" of Daubert: true peer review, editorial peer review, and scientific validity. NYUL Rev 1995;70: 100–134 at 118, 119.

[22]Jacobson RL. Scholars fault journals and college libraries in survey by council of learned societies. Chron of Higher Educ 1986; 32:1, 21.

[23]*See* Koren G. A simple way to improve the chances for acceptance of your scientific paper. New Eng J Med 1986;315:1298; Rothwell P, Martyn CN. Reproducibility of peer review in clinical neuroscience. Brain 2000;123:1964–1969 at 1968.

[24]Sales BD, Shuman DW. Experts in court: reconciling law, science, and professional knowledge. Washington, D.C.: American Psychological Association, 2005 at 59.

[25]Ashcroft R. Ethical Issues in Biomedical Publications. Bioethics 2003;17:371–374 at 372.

[26]Methods in Medical Ethics, Sugarman J, Sulmasy DP, eds. Washington, D.C.: Georgetown University Press, 2001 at 286.

[27]Andre J. Bioethics as practice. Chapel Hill and London: The University of North Carolina Press, 2002 at 57.

[28]Shuman DW, Sales BD. The impact of Daubert and its progeny on the admissibility of behavioral and social science evidence. Psychol Pub Pol'y & L 1999;5:3–15 at 7.

[29]Jonsen AR. Beating up bioethics. Hastings Cent Rep 2001;31:40–45 at 45.

8   Reliability of Bioethics Testimony

Experience

Not all bioethics testimony is based on generally accepted or peer-reviewed work. Some testimony is based on experience. This chapter considers when bioethics testimony that is based on an expert's experience is reliable. Because demonstrating reliability of experience-based testimony is more complex than demonstrating reliability by the Frye or peer review and publication criteria, the bioethicist who attempts it must be what Schoen calls a highly functioning practitioner—not only an active participant in the social situation of practice, but also a careful observer of and reflector on that practice.[1]

1. Bioethics Experience and Skills

Different kinds of bioethics work provide different experiences, and help develop different skills. To understand what a practicing bioethicist may be able to contribute, therefore, attorneys and judges may need to become familiar with a variety of types of bioethics work. Richard Ashcroft lists several, including "commissions of inquiry into new technologies or social problems, review boards to regulate research or access to technologies,

From: *Bioethics in Law*
By: B. J. Spielman © Humana Press Inc., Totowa, NJ

theoretical and empirical research, occasional and popular writing on topics of the day, consulting to public and private bodies, teaching of health care professionals and students, consulting on particular clinical cases, and drafting policies for hospitals, professional bodies or industrial organizations."[2]

In addition, the goals of bioethics work in particular contexts determine which kinds of skills are used. Encouraging bioethicists to consider the goals of bioethics work in various contexts and the relation of those goals to expertise, philosopher Scott Yoder writes:

> Ethics done in the clinic does not have exactly the same objectives as ethics done in the classroom or in the halls of government, nor does it require precisely the same skills. Thus to the extent that expertise is dependent on objectives, paying more careful attention to the context of ethical expertise should help to sharpen the debate. We may find it helpful to stop talking about ethics expertise in general and begin talking about ethics expertise in various domains. Close attention to the context in which the ethicist works and to the goals applicable in that context would vastly improve the discussion about appropriate indicators and measures of expertise.[3]

The context in which a bioethicist gains experience and the goals applicable in that context shape the kind of bioethics skills that he or she can rightfully claim in a legal context. Claims of "skill in clinical ethics" or "experience in teaching bioethics" are too general to match the needs of most legal cases.

Legal scholars Denbeaux and Risinger suggest that the literature of the group to which an expert belongs can be an important source for identifying the tasks and subtasks in which the group typically engages, and, therefore, a source for identifying a group's skills.[4] Such literature in bioethics is scarce. The American Society of Bioethics and Humanities report *Core*

Competencies of Health Care Ethics Consultation (ASBH Report) includes a list of broad competencies that clinical bioethicists should have.[5] (No comparable list exists for other kinds of bioethics work.) The ASBH Report's list of 12 skill areas and 9 knowledge areas could provide a starting place for specifying the skills that a clinical bioethicist might claim to have acquired through experience.

2. Skills and the Task at Hand

An ethics expert offering testimony must be prepared to demonstrate that the skill he or she claims to have used to develop the testimony is right for the particular legal issue. In the following excerpt of a Daubert hearing from *In re Diet Drugs*, an expert is challenged to demonstrate the reliability of his or her experience-based testimony (a Daubert hearing is a hearing within a trial, conducted before a judge, regarding the reliability of expert testimony).

In re Diet Drugs was a federal multidistrict case decided by the US District Court for the Eastern District of Pennsylvania.[6] Pondimin and Redux were diet drugs associated with primary pulmonary hypertension and heart valve damage. One of the contested issues was whether the pharmaceutical companies were required to include these health risks on the labels of diet drugs.

The plaintiff's attorneys wanted to use a medical ethicist to establish the pharmaceutical company's duty to disclose the risks. The medical ethicist, assuming that a duty could be based on a code of ethics, had claimed that he was an expert in interpreting health profession ethics codes. Based on the ASBH Report, it would not be surprising for a clinical bioethicist to claim some skill in interpreting codes of ethics relevant to clinical care of patients—if not to pharmaceutical codes.[7] In this excerpt from the Daubert hearing, the defense attorney is challenging whether the expert has the right skill for the particular legal issue.

Q.   Prior to your work in this litigation, you did not have any substantive experience applying these codes, by these codes, I'm talking about national pharmaceutical, Alliance, PHARMA, the IFPMA, the WHO?

A.   No, but let me explain that.

The Court: Did you or did you not, it is a relatively straightforward question, were you called upon to apply the standards in those codes?

The Witness: Not unlike a radiologist reading a chest X-ray, and medical ethicists deal with that all the time.

The Court: The answer is no.

A.   Fine.

Q.   You had never laid eyes on the World Health Organization code or applied that code, had you?

A.   I had laid eyes on it and I had not applied it.

Q.   You had never seen or applied the Pharmaceutical Alliance Code, had you?

A.   That's correct.

Q.   You had never applied or seen the code of the IFPMA or PHARMA code, have you?

A.   That is correct.

Q.   You had never seen a pharmacological company's code of conduct, you haven't seen AHPs?

A.   No.

Q.   You had never participated in the writing of the conduct for the pharmaceutical industry?

A.   I wished I had.

Q.   The answer is no?

A.   No.

Q.   There are people in those organizations whose job it is to interpret and apply those codes, correct?

A.   I don't know if that is the case or not.

Q.   In any event, prior to this litigation, you were not one of them?

. . .

Q.   You now testified that these are codes of ethics, correct?

A.   Some of them are, some of them are codes of marketing practice, they have the standard of code of ethics in this industry.

Q.   When you first were deposed in the Diet Drug Litigation, you didn't consider these to be codes of ethics, did you?

A.   I considered them to be codes of marketing practice, as I explained to you in one of our days of deposition, I have come to realize that codes of marketing practice within this industry constitute often a company's attempt at a code of ethics, like AHP's code is considered a code of conduct on its label.[8]

The defense attorney is probing several aspects of the fit of the bioethicist's skill to this case. His line of questioning implies that the task of reading health professionals' ethics codes is not the same as the task of interpreting pharmaceutical marketing codes. But how would a judge determine if this expert's skill in health care code interpretation could be transferred to pharmaceutical marketing codes?

The major technique for facilitating transferability in qualitative research is to provide a qualitatively rich and detailed description—a "thick description"—of the settings and contexts and subjects of previous work. If the expert could describe enough similarities between the contexts in which he had previously interpreted codes and the context of this case, a judge might be convinced that his skill transferred to this new context. To do this, however, a judge would need to know as much as possible about the original situation in which codes were interpreted as well as how they are similar to the case at hand.[9] If the expert in *Diet Drugs* had provided such a description, and the judge were persuaded that the expert's skill was transferable, he or she could then proceed to focus on the skills the expert actually used in this case.

3. Steps in Reaching a Conclusion

According to the Advisory Committee Note to FRE 702, if the witness is relying solely or primarily on experience, then the witness must explain how that experience leads to the conclusion reached, why that experience is a sufficient basis for the opinion, and how that experience is reliably applied to the facts. To explain how the experience leads to the conclusion reached, a bioethics expert must, according to the Note, be able to describe the skill of which the testimony is a product.[10]

The Supreme Court had insisted in *Kumho Tire Co.* v. *Carmichael* that simply characterizing the tire failure analyst's method as "a method of tire failure analysis which employs a visual/tactile inspection" was overly general for reliability purposes. Each step of the expert's four-part, two-factor test coupled with his assessment of the relative magnitude and significance of each of the four factors had to be identified.

In *Kumho Tire Co.* v. *Carmichael,* the Supreme Court expected the tire failure analyst to be able to describe in detail his four-part, two-factor test. Similarly, in *Diet Drugs,* the Federal District Court expected the bioethicist to describe in detail the steps by which he interpreted and applied the pharmaceutical code. The following excerpt shows a lost opportunity to articulate those steps.

By Mr. Waxman:

Q. Are these the kinds of codes that only medical and bioethicists can interpret?

A. No.

Q. Who else can interpret these codes?

A. I think anyone that reads and understands English.

Q. You don't have to be a doctor?

A. No, sir.

Q. Do you have to be a bioethicist?

A. No.

Q. If you can read and understand English, then your testimony is you can read and understand the codes that you referenced in exhibit 2, being a report.

A. In my opinion, yes.

This excerpt suggests that the expert's skill in code interpretation amounted only to the skill of reading English. However, if reading English were the only skill the bioethicist used to interpret the code, a judge would find he had no special expertise. In fact, even code interpretation based only on the code's text can involve complex interpretive rules or principles. Examples of such rules of interpretation in law are: "one part gives meaning to another part"; "expression of one thing means exclusion of another." Examples of rules of interpretation of professional codes can be found in the many aids to interpreting the American Medical Association's and the American Bar Association's codes—including their principles; opinions with annotations; comments; and ethics opinions. To the extent that the skill of pharmaceutical code interpretation involved textual skills other than simply reading the text, that skill might have been made more transparent to the court by articulating rules such as these.

It is apparent, however, that the expert thought that some of the skills used in code interpretation were not skills in working with texts, but skills developed through his experience as a physician and a clinical bioethicist. His frustration at the defense attorney's request to identify the connection between those experiences and his code interpretation is evident as the dialog continues:

Q.   Is that your testimony?

A.   Yes, it is because it seemed perfectly obvious to me that the sections in these codes were significantly violated and that as I explained a moment ago, as a medical ethicist, I am taught and trained to interpret codes of ethics, irrespective of where they come from, if they are health care codes. And I thought and still think it is clear that in fact these codes have been violated.

Now, I happen to also think that my approach to them being as straightforward as they are to me, is probably just in the context of my extensive experience [in] medicine and medical ethics.

The expert suggests at this point that the skill of code interpretation is not merely the skill of reading English. Rather, the skill is connected to "extensive experience in medicine and medical ethics." However, he is unable to articulate anything about the skill except the fact that he was trained in it. Identifying the precise pathways that generate testimony can be especially difficult when testimony is based primarily on experience, because experience can produce tacit knowledge:

> Some kinds of knowledge, sometimes collectively referred to as tacit knowledge, simply defy rational articulation—either because that knowledge relies on distinctions too subtle to be captured by existing vocabulary, or because it stems from a web of experiential and phenomenological correspondences too complex to be deduced into a rationalized verbal structure, or because it stems from socialized information that the possessor knows but is not consciously aware of knowing.[11]

Identifying the pathways that generate testimony may also be difficult in bioethics work because the rules for the work are underdeveloped.[12] Regardless of how challenging it may be for bioethicists to identify and describe the pathways by which they generate their testimony, however, Rule 702 requires judges to insist on it. The knowledge that judges use must be made explicit so that it can be transferred to others, understood by others, and evaluated by others—without reference to special intuition. Legal institutions cannot rely on experts' tacit knowledge, or on ways of knowing that are underdeveloped.

The bioethics expert in *Diet Drugs* did not only interpret the code ("The codes mean that nondisclosure of X is prohibited."). He also applied the code ("The set of circumstances in this case is a nondisclosure of X"). The taxonomic system used by those who apply codes is susceptible to description. Three examples of such taxonomic systems, offered by legal scholar D. Michael Risinger, in order of reliability, are the biological taxonomic system for animals, the *Diagnostic and Statistical Manual of Mental Disorders* (DSM), and the criteria for sufficiency of comparable real property sales for valuation purposes.[13]

Description of a taxonomic system for code application could include the rule or principle by which the expert and other code interpreters determine what counts as a "nondisclosure of X"; what counts as a nondisclosure of some other type; and what falls outside the category of nondisclosure altogether. Although judgment is necessary in using any system, the system itself must be made transparent, and, further, must not be entirely subjective.

Familiarity with such a taxonomy in law informs the defense attorney's cross-examination in the next section. He asks the expert about the rules and regulations for the reporting of adverse drug events to the FDA, and the requirements for drug labeling. However, the bioethics expert has no experience with those systems.[14]

4. Reliable Application of the Skill

Rule 702 requires not only that an expert be able to explain how the experience leads to the conclusion reached, but also how the experience is reliably applied to the facts.[15] In the following excerpt, the attorney focuses on the reliability of the expert's application of the code:

Q.   Prior to your work in this litigation, you the (sic) never discussed with any member of the organizations in promulgating these codes their interpretation or application of the code, correct?

A.   No, that is correct.

Q.   None of the opinions that you offered about how AHP violated these codes has been voiced by any of these organizations, correct?

A.   I don't know if that is the case or not.

Those who offer testimony interpreting and applying codes need not meet positivist criteria of reliability. To show that their testimony is not completely subjective, however, they might instead aim for "confirmability." This can be done by the use of a third party's assessment of the content of the work.[16] If the expert had replied that he had discussed his interpretation and application of the pharmaceutical code with a member of the promulgating organization, or consulted with experts in code interpretation and application outside the pharmaceutical field, he might have been able to provide information to help the judge decide that his interpretation and application of the code was reliable.

Code interpretation and application is only one set of skills that bioethics experts may claim to use in developing testimony, and it is not necessarily one that is used frequently. More frequently claimed, if only implicitly, is skill in historical reconstruction of interactions between physicians and patients regarding consenting to, or refusing, medical treatment or participation in research. We turn next to judicial reasoning regarding this bioethics skill.

5. Another Criterion and Other Skills

The criteria discussed in this chapter and in Chapters 6 and 7—experience, general acceptance, and peer review—are three criteria that might be useful in demonstrating or assessing the reliability of bioethics testimony. The advisory committee Notes to Rule 702 list other factors relevant in determining whether expert testimony is sufficiently reliable to be considered by the trier of fact, including what, for purposes of this section, will be referred to as the "fourth criterion":

> Whether the expert 'is being as careful as he would be in his regular professional work outside his paid litigation consulting'... [T]he expert [must] employ in the courtroom the same level of intellectual rigor that characterizes the practice of the expert in the relevant field.[17]

This criterion is important in demonstrating or assessing the reliability of strands of bioethics testimony that are not distinctively ethical. That is, if the strand is statistical, it should "employ… the same level of intellectual rigor that characterizes the practice of an expert in the…field" of statistics. And, importantly for historical reconstruction of consent conversations, if the strand is historical, it should "employ…the same level of intellectual rigor that characterizes the practice of an expert in the…field" of history.

But what if the strand is ethical, and based on bioethics experience? In the following example, an expert testified that informed consent was not properly implemented, based on a review of medical records and interviews with physicians.

Oddly, neither the skill of historical reconstruction of clinical events nor the skill of reviewing medical records is mentioned in the ASBH report. Nevertheless, these skills have been used to generate expert testimony not only in the case of *Heinrich ex rel. Heinrich v. Sweet*,[18] but in others, including *Stewart v. Cleveland Clinic Found.*, in which an expert bioethicists testified that informed consent to experimentation had not been obtained;[19] and in *Ramona Osgood, et al. v. Genesys Regional Medical Center*,[20] in which a bioethics expert testified that informed consent to life-sustaining dialysis had not been obtained.

The expert in *Heinrich* has expressed the opinion that informed consent was not given, based on his reading of the medical record and discussion with a physician. The attorney is challenging the expert on the third issue: Did you use the skill reliably?[21]

> Q. And what else did you—on what else did you form your opinion?

> A. I indicated, in my report, that I did not examine the full medical records of these two patients, but specifically asked Dr. Junck if he found any information about—

Mr. Doherty: Objection.

A. —about informed consent in those records.

Mr. Doherty: Move to strike.

The Court: No, I'm going to let that stand.

Q. In examining questions of informed consent, is the inquiry you made of Dr. Junck the type of inquiry which you reasonably rely upon in the field of medical ethics in forming opinions or inferences upon the subject of medical ethics?

Mr. Doherty: Objection.

Q. That is, to discuss it with a medical doctor?

Mr. Doherty: Standard of medical ethics.

The Court: Well, let me try it this way: You knew that you were going to be called here to talk about the issue of informed consent, an issue in this case? You knew that from the first time anyone contacted you; is that right?

The Witness: Yes.

The Court: Now, the whole issue of informed consent, you've testified to, is one of those things about which you teach and indeed conduct research; that's right?

The Witness: Yes.

The Court: Okay. Now, in forming your opinions to teach, are you content to rely upon the examination of records by another doctor, not a treating doctor, but another doctor who said he examined the records? Are you content to rely upon what he says is found in the records or not found, in reaching your opinions about what those records show? Is that how you do it in doing your research?

The Witness: In clinical consultation, where we're advising clinicians, advising physicians about matters such as informed consent, we rely not only on the public and professional documents that have already been mentioned, but necessarily those, say, participating in an ethics consultation will rely on the review of others, including physicians, in determining whether information has been provided in an adequate way.

The Court: Okay. But, as I follow this here, in this case, we don't—you didn't talk to anyone who was involved in the procedures or treatments or whatever we call them, did you? Right? Back in 1960 and '61?

The Witness: No, no.

The Court: All right. But you did talk to Dr. Junck?

The Witness: Yes. And Dr. Grodin, as well, yes.

The Court: All right. You talked to two doctors, but neither of them are treating doctors; correct?

The Witness: That's correct.

The Court: And my question to you is: Is that how you do your research? Are you content in your research and content in your teaching, in either consulting about ethical matters or in teaching and forming your opinions, to talk to doctors who say they have reviewed the records, but aren't themselves treating physicians, and get from them their information about what's in or not in the records? I'm just asking, is that how you do it generally?

The Witness: We use different methods for different problems that need to be addressed. Given that even in a simple case, there may be thousands of pages, often we have to rely on the consultation with others who have reviewed the record.

So this is not an unusual situation where I would rely on, in this case two physicians who have looked at medical records or depositions, to give me assurance that there was nothing, other than in the documents that I'd already seen, that would suggest that there was any evidence that the consent had been obtained.

The Court: Uhm-hmm. And not just you, but there are other professors, other teachers of medical ethics, are there not?

The Witness: Yes.

The Court: Are you familiar with the standards that they follow in conducting such research and consultation?

The Witness: Oh, I'm—I've met with over a hundred ethics committees, and the procedure I describe is exactly the way an ethics consultant would handle the case.

The Court: All right.

The Witness: If there was ambiguity, we'd look at some of the documents, but there's no way that for every consultation we can look at every piece of paper in every medical record.[22]

The attorney and judge may have been trying to ask a question about reliability. Perhaps they had in mind the fourth criterion. However, notice that the judge and expert jumped from the experience of teaching to the experience of research to the experience of clinical consulting. Neither is clear regarding the logically previous questions, "Does the expert have the right experience and skills for the task at hand?" and "What steps were used to produce the testimony?" Having failed to obtain precise answers to these questions, the judge seems to have been left with the impression that ethics consultants and committees are in the business of historically reconstructing informed consent to 1960s research, and that teaching about informed consent involves review of medical records and interviewing nontreating physicians.

Even if questions about possessing the skill and steps in generating testimony are clearly asked and answered, application of the fourth of the Advisory Committee's criteria is complicated if bioethics testimony is being offered. Confusion arises precisely because bioethics testimony can have so many strands, and be generated from so many different methods and experiences in different work contexts that have different goals. What everyone involved in this case failed to clarify regarding the bioethicists' work experience is Yoder's point, highlighted early in this chapter: ethics consulting work has its own objectives, as do bioethics research and bioethics teaching. Ethics consultants review the medical records and interview physicians, as do some ethics committees, but not ordinarily for the purpose of historical reconstruction. Even if an ethics committee does engage in case review, it will not be a review of a case that is decades old. The reason a consultant or committee will look at medical records or interview physicians is usually to help as a decision regarding patient care is being made or modified. Ethics committee and consulting experience would, therefore, neither require nor help develop the skill of historically reconstructing long-past clinical events.

Bioethics research experience can take a variety of forms. Regardless of the form, however, if the research question required a medical record review and physician interviews, the selection process for both records and interview subjects would have to be more carefully justified than they were here. The experience of teaching is one many bioethicists possess, but the skills involved in teaching do not overlap significantly with those required to develop reliable testimony for this case.

6. Summary

Using examples from *In re Diet Drugs* and from *Heinrich ex rel. Heinrich* v. *Sweet*, this chapter illustrated that demonstrating the reliability of experience-based bioethics testimony is not as straightforward as offering a list of one's experiences; but neither is it impossible. The task is complicated by the many hats that bioethicists wear, and by the variety of their work experiences and objectives. In summary, the bioethics experience must be an experience that requires skills appropriate for the legal task at hand; the steps of the skill that lead to the expert's conclusion should be identified; and to demonstrate reliability, quality of information procedures should be used and articulated. Law is as open to experience-based testimony as it is to other bioethics testimony, when experts can demonstrate its reliability.[23]

Endnotes

[1]Schoen D. The reflective practitioner: How professionals think in action. New York, NY: Basic Books, 1983.

[2]Ashcroft RE. Bioethics and conflicts of interest. Stud Histo Phil Biol & Biomed Sci 2004;35:155–165 at 156. The forms of bioethics work are so varied that Carl Elliot has characterized bioethicists as a "strange hybrid of policymaker, pundit, and bureaucrat." Elliott C. Throwing a bone to the watchdog. Hastings Cent Rep 2001;31:9–12.

[3]Yoder SD. The nature of ethical expertise. Hastings Cent Rep 1998;28:11–19.

[4]Denbeaux MP, Risinger DM. Kumho Tire and Expert Reliability: How the Question You Ask Gives the Answer You Get. Seton Hall L Rev 2003;34:15–75.

[5]Society for Health and Human Values—Task Force on Standards for Bioethics Consultation. Core competencies for health care ethics consultation. Glenview, IL: American Society for Bioethics and Humanities, 1998 at 13–15, 20–21.

[6]*In re*: Diet Drugs (Phentermine, Fenfluramine, Dexfenfluramine) Products Liability Litigation; This Document Relates To All Cases, 2001 U.S. Dist. LEXIS 1174 (2001).

[7]Society for Health and Human Values—Task Force on Standards for Bioethics Consultation. Core competencies for health care ethics consultation. Glenview, IL: American Society for Bioethics and Humanities, 1998 at 20–21. Reading codes is described as "knowledge" rather than a "skill" area in the ASBH Core competencies

report. Relevant codes of ethics, professional conduct, and guidelines of accrediting organizations as they relate to ethics consultation is one of the nine knowledge areas. "According to the
report, for basic knowledge in this area, one should read the relevant code or manuals.... In order for advanced knowledge in this
area to be available to the process of consultation, ethics consultants should know who the contact persons might be to discuss the
area in question (e.g., the person or persons responsible for the
JCAHO survey). They should also know where to find the code
or accreditation manual."

[8]*In re* Diet Drugs Prods. Liab. Litig., deposition of John Joseph La
Puma on December 5, 2000, at p. 224 line 4 to p. 225 line 15, and
p. 226 lines 6–20.

[9]Fishman DB. The case for pragmatic psychology. New York: New
York University Press, 1999 at 185.

[10]Nothing in this amendment is intended to suggest that experience
alone—or experience in conjunction with other knowledge, skill,
training, or education—may not provide a sufficient foundation
for expert testimony. To the contrary, the text of Rule 702
expressly contemplates that an expert may be qualified on the
basis of experience. In certain fields, experience is the predomi-
nant, if not sole, basis for a great deal of reliable expert testimony.
See, e.g., United States of America, Plaintiff–Appellee, v.
Kathleen Kremser Jones, Defendant–Appellant, 107 F.3d 1147
(1997) (no abuse of discretion in admitting the testimony of a
handwriting examiner who had years of practical experience and
extensive training, and who explained his methodology in detail);
Henry Tassin, et al. v. Sears, Roebuck and Co., et al., 946 F.Supp.
1241, 1248 (1996) (design engineer's testimony can be admissible when the expert's opinions "are based on facts, a reasonable
investigation, and traditional technical/mechanical expertise, and
he provides a reasonable link between the information and procedures he uses and the conclusions he reaches"). See also Kumho
Tire Co. v. Carmichael, 526 U.S. 137, 155 (1999) (stating that
"no one denies that an expert might draw a conclusion from a set
of observations based on extensive and specialized experience.").

If the witness is relying solely or primarily on experience,
then the witness must explain how that experience leads to the

conclusion reached, why that experience is a sufficient basis for the opinion, and how that experience is reliably applied to the facts. The trial court's gatekeeping function requires more than simply "taking the expert's word for it." See Daubert v. Merrell Dow Pharmaceuticals, 43 F.3d 1311, 1319 (1995). ("We've been presented with only the experts' qualifications, their conclusions and their assurances of reliability. Under Daubert, that's not enough.") The more subjective and controversial the expert's inquiry, the more likely the testimony should be excluded as unreliable. See James R. O'Conner, Plaintiff–Appellant, v. Commonwealth Edison Company and London Nuclear Services, Inc., Defendants–Appellees, and United States of America, Intervenor–Appellee, 13 F.3d 1090 (1994) (expert testimony based on a completely subjective methodology held properly excluded). See also Kumho Tire Co. v. Carmichael, 526 U.S. 137, 1551 (1999) "(I)t will at times be useful to ask even of a witness whose expertise is based purely on experience, say, a perfume tester able to distinguish among 140 odors at a sniff, whether his preparation is of a kind that others in the field would recognize as acceptable."

[11]Dowdle MW. II Jurisprudential considerations: Deconstructing Graeme: Observations on "pragmatic psychology," forensics, and the institutional epistemology of the courts. Psychol Pub Pol'y & Law 2003;9:301–332 at 312–313.

[12]Agich G. The question of method in ethics consultation. Amer J Bioethics 2001;1:31–41 at 37.

[13]Risinger DM. Preliminary thoughts on a functional taxonomy of expertise for the Post-Kumho World. Seton Hall L Rev 2001;31:507–537 at 524. I am indebted to Risinger's discussion of translational expertise at 517–526.

[14]*In re*: Diet Drugs transcript, cross examination of John Joseph La Puma on December 5, 2000, at 203–214.

[15]USCS Federal Rules of Evidence. 702 Rev. 2002.

[16]Fishman DB. The case for pragmatic psychology. New York: New York University Press, 1999 at 186; Denbeaux MP, Risinger DM. Kumho tire and expert reliability: How the question you ask gives the answer you get. Seton Hall L Rev 2003; 34:15–75 at 57.

[17]Sheehan v. Daily Racing Form, Inc., 104 F.3d 940, 942 (7th Cir. 1997). See Kumho Tire Co. v. Carmichael, 119 S.Ct. 1167, 1176 (1999). Daubert v. Merrell Dow Pharmaceuticals, Inc. requires the trial court to assure itself that the expert "employs in the courtroom the same level of intellectual rigor that characterizes the practice of an expert in the relevant field." Advisory Committee Note, F.R.E. 702.

[18]Heinrich v. Sweet, deposition of Robert M. Veatch on September 23, 1999, at p. 49 line 12 to p. 52 line 25.

[19]Jean K. Stewart, Executrix of the Estate of Daniel V. Klais, et al., Plaintiff–Appellants, v. Cleveland Clinic Foundation, et al., Defendants–Appellees.

[20]Ramona Osgood, et al. v. Genesys Regional Medical Center, Deposition of Carl Cohen, Decided by Circuit Court, Genesee County, Michigan, Case No. 94-26731-NH, February 16, 1996, p. 6 line 5 to p. 49 line 21.

[21]Heinrich v. Sweet, deposition of Robert M. Veatch on September 23, 1999, at p. 49 line 12 to p. 52 line 25.

[22]The attorney may also have been trying to have the testimony excluded as hearsay, or included via a hearsay exception.

[23]Fishman DB. The case for pragmatic psychology. New York: New York University Press, 1999, at 170.

Conclusion

Prospects for the Future

If bioethics and law are to collaborate more effectively in the future, each will need a solid understanding of how and why bioethics is used—or not used—in law. By analyzing, on a case-by-case basis, the interactions in which bioethics has come to law during the last decade, this book illustrates a line of interdisciplinary inquiry intended to advance that understanding.

During the next decade, direct absorption of bioethics material should not be expected; nonetheless, each of the bioethics communications around which the chapters of this book are organized will be found in new law. Health care ethics committee determinations and institutional review board determinations will continue to be relied on if they do not contravene core legal norms. The determinations will continue to be rejected and used as negative exemplars, however, if they override individual rights and procedural norms. Judges will use bioethics commission reports both representationally and rhetorically, in ways their authors intended and in ways they could scarcely have imagined. They will continue to do so without regard for the bioethics norm of consensus. Subpoenas for empirical bioethics research will be rare, and subpoenas for non-empirical bioethics scholarship rarer still, but bioethicists will learn what is negotiable during discovery and what is not. Expert testimony will contribute to judicial

From: *Bioethics in Law*
By: B. J. Spielman © Humana Press Inc., Totowa, NJ

thinking on a wider range of issues than health care ethics committees, institutional review boards, or commissions can. However, expert bioethicists, the attorneys who work with them, and judges will need to learn how to better assess reliability. Critically assessing one's own methodologically eclectic work— in other words, becoming a reflective bioethics practitioner—is indispensable for bioethicists involved in this work.

The goal, of course, will not be to bring bioethics reasoning in line with legal reasoning. Legal and bioethical reasoning will, one hopes, remain distinct, continuing to operate alongside each other and to influence each other. The grounds on which judges respond to bioethics will continue to be legal, not bioethical. Judges will incrementally make use of bioethics norms, if they enhance or supplement law's core norms. However, judges will block uses of bioethics that would erode or supplant those norms. A reasonable goal for future interactions, therefore, is to avoid direct clashes between bioethics and core legal norms that prevent law from recognizing a bioethics communication as potentially helpful.

All of this assumes that core legal norms, and the role of the judiciary in interpreting them, will not be destroyed by other means. Needless to say, under either the "unitary executive" or the Christian fundamentalist theocracy now being promoted in the United States, the line of scholarship illustrated here would not be necessary. Under such a regime, an executive could completely open law to norms approved by theocrats. Law and bioethics would continue to collaborate, but instead of developing interactions that are more complex, as one would expect of two evolving systems, the patterns of their interactions would be simplified as society regressed. Assuming that core legal norms and the role of the judiciary are stabilized, however, opportunities to struggle with the challenges of bioethics in law, as I have done in this volume, will continue to grow.

Index